LEVELS OF
AWARENESS

LEVELS OF AWARENESS

From Microbes to Humans

Vern A.
Westfall

LEVELS OF AWARENESS
FROM MICROBES TO HUMANS

iUniverse books may be ordered through booksellers or by contacting:

iUniverse
1663 Liberty Drive
Bloomington, IN 47403
www.iuniverse.com
1-800-Authors (1-800-288-4677)

ISBN: 978-1-5320-1509-0 (sc)
ISBN: 978-1-5320-1511-3 (hc)
ISBN: 978-1-5320-1510-6 (e)

Library of Congress Control Number: 2017900597

Print information available on the last page.

iUniverse rev. date: 01/24/2017

INTRODUCTION

It has taken us a very long time to rediscover what we knew instinctively before we became civilized. Having traced life's development from the diversity of extinct creatures to extant forms we have reconnected with our past and the evidence is clear. All living things, including humans, are related. Complex languages and conceptual advantages set us apart, but do not separate us from life's adaptive chain of development.

We have extended our examinations and opened our minds to accept our natural place, but semantic constraints remain and slow the development of new perspectives, especially those associated with the awakened state of matter. Words, like cognizance, intelligence, and reason, carry subliminal egocentric connotations that limit broad comparative assessments and focus our assessment of awareness on ourselves instead of on its presence in all living things.

From early single cells to the mass of neural cells in the human brain, the history of life has a common trace with awareness as an essential element. To appreciate the awakened state of matter as more than animated organic molecules, we need to search beyond our egocentricity and examine the evolution of awareness as it developed in concert with life's physical attributes.

By expanding our limited definitions of awareness to encompass a broader perspective, we can begin to trace its evolution, define its many attributes, and understand its importance in natural selection.

By examining how nature produces emergent properties, how ordinary matter becomes self-energizing, and how complex organic molecules become capable of replication, we can begin to understand how ordinary matter becomes both alive and aware.

By examining the essential evolutionary partnership between awareness and physical forms, we can begin to trace the expansion of awareness in scope and acuity and develop a self-image more appropriate to our evolved aware state.

TABLE OF CONTENTS

CHAPTER ONE

Chapter Contents

NATURE'S CREATIVE PATTERNS

Perceptions

Before science, perceptions explaining life came from ancient tales and myths. Our current perspectives form differently. Current perspectives are the result of discoveries made using cooperative investigative efforts and sensory enhancing tools. With the help of our inventions, we are now able to look outward into space and backward in time and find ourselves, not at the center of a small well-planned universe, but on a small planet orbiting a yellow star near the edge of a spiral galaxy.

Using our sensory enhancing tools, we are able to look deep into the workings of the smallest elements of matter and find that we, and everything else in the universe, are composed of the same atomic particles. We have deciphered many of the joining rules for these particles and are beginning to understand how their combinations create the emergent properties of gravity and magnetism, and how in turn, these emerged energy states gather the materials that gave them birth to create stars, planets, life, and awareness.

At first glance, nature's creative methods appear to be a disconnected series of events rather than a continuous creative process. We overlook the continuous flow of creation because nature has programmed us to look at and understand our surroundings piece by piece and one event at a time. To resolve our perceptive disconnect with nature's methods we locate individual creative events that fit comfortably into our neural programming, reduce complexity by using symbolism, create descriptive and predictive models using mathematics, and use analogy to create conceptually acceptable larger images,

The theories we create to understand the world around us are limited reflections of reality, but nature used the same method to create our awareness that it used to create the reality we are attempting to understand. Our mental capacities, like our physical bodies, have been formed by nature's creative method of exposing matter and joining rules, to energy states that induce it to join and in turn join again to create a hierarchy of more complex materials and more emergent energy states.

Our complex neural capacities were formed by nature, compliment nature, and allow us to create conceptual groupings "in-sync" with the way the universe is organized. Unfortunately, our advanced state of self-awareness evolved before we developed the ability to examine ourselves' critically. As a result, our early self-awareness distorted and magnified our self-image. Before we began using scientific methods, our self-awareness produced only egocentric results. Early revelations and assumptions concerning our place and purpose reflect these self-aggrandizing perspectives and continue to separate us from the reality around us. However, as we learn to read the book of nature, our egocentricity is slowly being set aside. As we learn more the comfort provided by a human centered universe is being replaced by an impersonal universe where humans seem to be of little consequence. Our advanced state of awareness is creating a dilemma. Old perspectives give humans significance but seem inadequate. New perspectives seem relevant but make humans insignificant. Unwilling to accept insignificance, we cling to old perspectives while continuing to search for purpose in nature's processes

As nature's patterns are reveled, so are nature's methods. One of these methods is natural selection, and as we examine nature's random method of molding life and awareness, we remain hopeful that we will discover significance, a sense of purpose and a way to combine recent discoveries with more comforting perspectives.

Older perspectives describe nature in terms of human needs and behavior. Scientific perspectives deal with the universe differently. Science examines particles 10 to the 25th power smaller than man and explores space 10 to the 27th power larger than man. We find ourselves miniscule in an unimaginably large universe and yet immense compared to its basic constituents. We find our span of awareness to be but a blink between the start of the universe and an equally long time before its projected end. We are near the center of size and time, staring at the universe around us in amazement, exploring it in detail, questioning its purpose, and listening to the echo of our own self-awareness.

As we explore, we are discovering that we are as much a part of the universe as the galaxies around us, and are composed of the same atomic particles. We have been created and shaped by forces and patterns we do not fully understand for a purpose we can't find and which may not exist. Living things are the stuff of the universe evolving, an awakening dust gathered into complex forms struggling to understand their surroundings, and themselves'. The significance of stardust becoming aware is unknown, but if awareness is relevant, it may give all living things purpose and human awareness a special significance.

Patterns

Our examinations of nature's materials, forces and creations have disclosed details beyond counting. To keep track of what nature is telling us we have become specialists with ever narrowing perspectives, each specialist capable of reading only a few pages of nature's continuing disclosures. The pages of discovered information are most valuable when connected. When everything has been discovered and connected, the information will form a complete volume, a book of the universe as interpreted by a group of tiny living creatures on a tiny planet. This is, of course impossible.

We will never come close to discovering all of nature's details. We can however look for obvious patterns that repeat as the universe creates more details and use the patterns as guides for further inquiry.

Cosmological Patterns

The patterns we observe at the largest of scales we group into a grand concept called cosmology. These largest of patterns include the spontaneous emergence of all matter and energy from a singularity followed by the mutual destruction of matter and anti matter and a diffuse cooling as the new universe expanded. We infer a non-uniform distribution of energy due to some unknown initial influence and a shift, early in the creative process, from a dark opaque universe to one in which light escaped its plasmic confines. As the universe cooled, gravity and magnetism emerged and began gathering matter, into immense sinuous clouds, proto stars and galaxies. The first giant stars concentrated gravity compressing and fusing hydrogen and helium into carbon, oxygen and silicon. As matter gathered, elliptical orbits and spheres emerged as natural states of motion and form and the early universe began to take shape, but the creative crush of early giant stars was insufficient. Another more powerful creative step was needed to create the additional elements needed for creation to continue.

Early giant stars remained stable until they fused lighter elements into iron. The fusing of iron produces insufficient energy to balance a large stars gravity and the moment iron appears, the star's core collapses and blows off its outer shell as a super nova. The compressive force of a super nova fuses the additional elements needed for life and adds them to the mix of interstellar-stellar gas and subsequently to second-generation, smaller, longer-lived stars and their planets.

Cosmological patterns are an evolving series of interdependent steps with atoms gathering to produce the emergent forces needed to form the universe's macro features.

Atomic and Molecular Patterns

We describe the constituents of matter according to their characteristics and combining properties and divide our micro discoveries into two areas of study, particle physics and chemistry. We have identified the basic constituents of atoms and the joining patterns of their constituents. We have identified the natural atoms, have created a few of our own, and have studied the ordered combinations of atoms to form molecules. We have extended our understanding of joining rules to create a grand concept for organic molecules and have identified the carbon based congregates that make up life.

Complexity and diversity increase exponentially at every level. Quarks combine in a limited number of ways to produce protons and neutrons. Electrons combine with protons and neutrons with more latitude to produce a hundred plus natural atoms, and atoms combine with each other in many more ways to produce an extreme variety of chemical elements that in turn combine to make up an extreme variety of organic and inorganic materials.

Organic Patterns of Development

To understand living forms we group the patterns we observe into a grand concept we call biology, and to accommodate our conceptual limitations, we divide the totality of living forms into kingdoms, phyla and species. We use these concepts to trace life's origins and its evolution, and to decipher the chemical codes that direct and control the process of life's development. The patterns for life have many interdependent relationships influencing its development and its survival.

- Environmental restrictions
- Adaptability
- Symbiotic relationships
- Parasitic relationships
- Predatory effects

- Competitive influences
- Cooperative influences
- Chance/Circumstance

Life sciences now dominate much of our scientific efforts and reflect our continuing need to answer the universal questions imposed by our self-awareness.

Patterns of Awareness

Still missing from our basic set of explanatory categories are explanations for the combinational and developmental patterns of awareness itself. Initial efforts are starting to identify the genes responsible for organs of awareness and to trace their origins in many species. Research is also beginning to link genetics and behavior, but questions as to how awareness springs from a physical form, or how the elements of awareness have evolved, and how the evolution of awareness relates to natural genetic selections, remain unanswered. We have not grouped patterns of awareness into a grand concept because scientific methods are best suited to the examination of nature's physical aspects.

Attempts to create hypothetical explanations for the development of awareness can also be easily misdirected. In the 1870s, Darwin touched on this dilemma, in "The Expression of Emotions in Man and Animals", when he expressed his disapproval of attempts to link human psychology to evolutionary principles in a field of study called "Sociobiology". A pseudo-scientific study that quickly became a platform for sensationalism by popular science writers.

Awareness, as a concept, may be beyond our current *inductive* scientific methods but, *(If we are careful not to stray from a serious examination into an egocentric analysis),* it is fair game for deductive examination and philosophical inquiry.

Evolutionary Theory and Genetics

Early Observations

Early man observed and understood the basics of animal breeding while he was still subsisting as a hunter-gatherer. The observable fact that there is a slow succession of changes in physical characteristics from generation to generation in both plants and animals has never been questioned. A slow evolution of characteristics in living things is obvious and is universally accepted. Also accepted is the fact that the slow process of generational change is occasionally interrupted by the advent of an individual with physical or behavioral characteristics very different from their ancestral stream. These anomalies have been noted for eons and until recently, have been interpreted as super-natural occurrences or interventions by an unseen god.

The idea of evolution is much older than Darwin's writings. The concept remained buried because there was no observable mechanism to explain how nature could practice selective breeding using random methods, or how a small errant modification could lead to a completely new variety or species. A purposeful creation by the gods with a unique pre-existing design for each creature has been the accepted explanation throughout much of man's history. In spite of overwhelming evidence, this view continues to be accepted by many.

Evolutionary theory lay dormant for many centuries but continued to tease the observant by prompting the same inevitable questions. With the discovery and exploration of the New World, hundreds of new life forms were discovered. New plants, new insects, new birds, new fish and new animals were observed. As sketches and samples of these amazing plants and animals arrived in the Old World, questions as to their origins were inevitable and the pursuit of an overall explanation for the diversity of living forms was revived. Why did functional forms seem to fit their environment, were long necks and strong jaws developed out of need, and how were these traits passed on to later generations? How could stretching one's neck transform the necks of the next generation?

Charles Darwin's grandfather, Erasmus, struggled with these age-old questions long before Charles took up the quest. On the surface, it seemed obvious. If an animal needed a long neck to reach food, it grew one and then passed the long neck quality on to its offspring, but how?

Man discovered early that containing and controlling selected animals was much more efficient than following their migrations. Animal domestication predates civilization and man learned quickly that he could change the characteristics of an entire heard by selective breeding, and as he directed the inherited characteristics of his herds, he could not help but wonder if similar changes in wild animals might be due to selective breeding by some unseen hand. Early man asked the right questions, but was far from being able to answer them. It took Charles Darwin's observations of birds, plants and animals from various islands in the Galapagos to refine the questions and create a relevant answer, not a final answer, but one that put us on the right path.

Darwin noted that identical species on different islands exhibited slightly different characteristics in size and color. If there was no apparent difference in their environments, where did the differences come from? The obvious answer was that the changes were random and the results became preferred for insignificant reasons. What Darwin was observing and recording, (but not explaining), was genetic drift. Equally important, were his observations of adaptive changes in a species when slightly different environments were presented. Separated groups originating from identical stock could apparently have their physical characteristics selected to fit a changing environment when nature practiced its own version of selective breeding. The difference between nature's selective process and man's selective process is in man's ability to reach his selective breeding goals in a few dozen generations while nature takes hundreds or thousands of generations and, unlike man, has no final goal in mind, and depends upon genetic errors to offer up divergent characteristics.

Natural selection requires a very large number of individuals and extremely long periods to affect changes. The process is completely random and has no stopping point. A very slight modification in form or function that gives a slight advantage in breeding or survival will carry forward as a small statistical preference that may occasionally, become a strongly preferred trait. A new trait, or physical attribute, arriving spontaneously as a genetic mutation faces the same test. The new anomaly must add a small statistical mating or survival advantage immediately, and then remain an advantage over the long term.

Advantageous traits, arriving from the slow drift of individual characteristics, or an occasional spontaneous modification, are both subject to cancellation or enhancement if the environment that supports them changes. The process of natural adaptation is slow. Environmental changes however, can be swift. A trait that has a strong survival advantage can be cancelled in just a few generations by an environmental change or by the introduction of a new predator. Survival and mating advantages are also a complex mix of causes and conditions. They are never a one at a time test and a mix of preferred and detrimental characteristics can carry poor performance traits forward through many generations. The natural selection process is random, has no apparent goal, and yet it has created an amazing array of divergent life forms ranging from protozoa to mammals. The process of genetic drift and environmental testing is effective and continues without any apparent stopping point.

We are so accustomed to our own creative methods that we have difficulty imagining any other. For most of human history, man insisted on the preposterous notion that nature, or the creator, had to do things like humans and within periods humans could comprehend. For us a pocket watch is always evidence for a watchmaker. We assume that a design and a purpose for the watch necessarily preceded its creation and, that when the watch is complete, it will stop being modified. Fortunately, nature doesn't do things like humans. In nature, basic laws govern firmly but allow extreme latitude. An advanced design is not needed to create a planet a mountain or a living thing. Creative events can occur very quickly, too fast for a human to notice, or very slowly, too slow for a human to comprehend, and the creative process is ongoing, nothing is ever completed, and as a result, life persists.

The creator, if one exists, appears to be playing an unending game of dice, continually casting the die into every environment and whenever doubles come up, lets them reproduce until a change in the environment cancels the play. From this perspective, life has no special significance and no purpose beyond the game itself. There is however, another perspective that may give the observable creative process significance. Continual play and adaptive changes both testify to a natural tendency to fit life into every possible environment and to adjust living forms and awareness as needed to maintain a living presence as environments change. Life seems so important that it is tested against every possibility, adjusted as needed to persist, incorporates increases in complexity as a survival advantage, and is randomly introduced and transforms itself to fit into every possible physical domain

To insure life's continuation it never stops adapting. Explaining this persistence may be beyond our capability, but denying it is equally difficult. The development and continuance of life appears to be as much a part of the universe as the laws of physics. The hierarchy of physical creations seems echoed in the hierarchy of living creations. From atoms to organic molecules and from nucleic acids to cells to multi cellular forms, life is ordered in a hierarchy that is obvious but difficult to explain.

The universe seeks entropy but is opposed by the creative force of gravity and by the persistence of life as it gathers organic molecules into complex energy forms. This opposition to entropy creates sequences of ordered events that we could interpret as creative intent, but we have no evidence to support such a conclusion. We can only observe and continue to ask, does life have any purpose beyond being a persistent part of a universal mechanical mix?

Our self-awareness makes our questions inevitable. Why am I here? What am I supposed to be doing and what happens when my awareness ends? Aware of our own existence we have no choice but to answer that our existence must have a purpose and that we are important in the scheme of things, otherwise there is no point to the inquiry. To assure ourselves that we have significance we have been very inventive. Creation theories, myths, religions and scientific inquiry, all testify to our need to find answers to these universal questions

Nature's book is open for us to read, but the conclusions we draw will always be self-aggrandizing. Getting beyond our own limited self-awareness will always be difficult. We seem doomed to read nature's book with our own reflections blurring the pages. We are very small creatures in a very large universe and are forced to observe our surroundings at our own living pace, within our own limited scope of awareness, and restricted by our conceptual limitations.

Evolution, the Concept

Natural selection is the filter used by nature to guide living organisms to a successful fit with their environment. Darwin called evolution 'natural' because it appears to be undirected and takes place without intervention. A river flowing downhill is a natural occurrence and life adapting to its environment is just as natural. We are more comfortable with the natural process of water flowing downhill than with the natural selection of living forms because, water flowing downhill is an easy observation and fits neatly into our pace of awareness. The downhill water concept takes only a second to observe, involves only a simple subject and requires very little conceptualization. The evolution of living forms, on the other hand, takes place at a pace well below our observational rate, involves an extremely complex set of observational subjects, and requires considerable conceptualization.

Religious explanations for the creation of various species often conflict with scientific findings. Paradoxically, genetics, as a scientific discipline, began in an Augustinian monastery where a monk; Gregor Mendel, used careful observation and statistical analysis of pea plants to produce evidence that hereditary characteristics are passed from generation to generation as distinct particulate factors. His paper, published in 1866, (seven years after Darwin published his "Origin of Species"), was a masterly use of scientific method. Both Mendel's work and Darwin's ideas remained unappreciated for years. Mendel's work remained an unread paper for 31 years. During the same period, Darwin's conclusions were embroiled in a lengthy confrontation with alternate explanations based upon creationist theories and the idea that organs and other living characteristics evolved or were discarded based upon their use or disuse. These ideas remained viable alternatives until Mendel's papers were rediscovered in 1900. The rediscovery of Mendel's work revitalized the idea that heredity could be passed by discrete units and that it could be the basis for evolutionary changes. Mendel's observations and conclusions gave new support to Darwin's point of view. The controversy between mutationists, (Mendalians), and biometricians, (Lemarkians), approached resolution in the 1920s when mathematical arguments showed that variations in living characteristics could be explained by Mendel's laws and that small variations could become cumulative effects and result in major evolutionary changes, as suggested by Darwin.

The fact that Darwinian ideas became dominant because of scientific observations made by a religious monk, should help those who see science as the enemy of religion, understand that science is simply a careful examination of our surroundings, followed by experimentation to check the validity of any conclusions. When nature tells us something different from our religious teachings, we have to choose or seek a compromise. The theory of evolution is the result of many compromises, some by science as new evidence required the modification of old theories, and some by religious thinkers as simple explanations became inadequate.

St. Augustine allowed that some life forms must have developed after the great biblical flood since Noah's ark could not have been large enough to hold all observable contemporary forms. Theologians of the middle ages, like St. Thomas Aquinas, also accepted the possibility that living things could be generated from inanimate matter, and that such a possibility was not incompatible with religious teaching.

Darwin's Observations

Darwin returned home after his voyages on the H.M.S. Adventure and the H.M.S. Beagle in 1837 and began to deal with the question of the origin of species. In 1844, he collected his notes into a sketch of conclusions that he thought probable and began work on a manuscript. He continued to develop his manuscript until 1859. Darwin anticipated that it would take many more years to complete his work, but he published his abstract early, influenced by the fact that Alfred Wallace, studying the natural history of the Malay Archipelago, had arrived at almost the same general conclusions and was ready to publish. Darwin's work was published in 1859.

Darwin formulated his conclusions about natural selection early but took twenty-two years to write and publish "The Origin of Species". He knew the theory would impact Victorian attitudes adversely and was especially sensitive to the opinions of his Unitarian wife. He was also cautious because, although he was confident in his observations and conclusions, he could not explain the biological mechanisms involved. It took another forty years before Mendel's papers were rediscovered and added verification to Darwin's conclusions. It took another fifty-three years before Crick and Thompson deduced the molecular structure of the gene and only after the discovery of nucleic acids and the genetic language of DNA, was it possible to put all the pieces together.

Darwin's work, "The Origin of Species", follows a logical order as outlined in the books table of contents. The table begins with observations of variations in living forms under domestication followed by observations of variations induced by nature including:

- The struggle for existence,
- natural selection,
- laws of variation,
- difficulties and objections to the theory of natural selection,
- the natural selection of instincts, reflexes and habits,
- hybridism,
- imperfections in the geological record,
- the geological succession of organic beings

- geographical distributions,
- mutual affinities,
- Morphology, embryology, and rudimentary organs and a final chapter; recapitulation and conclusion.

His work is filled with details from his observations and with excerpts from other naturalists and observers. The explanations and, sometimes hesitant, conclusions within his work point toward natural selection, but the introduction to his work makes it clear that he was committed to the idea:

"I can entertain no doubt, after the most deliberate study and dispassionate judgment of which I am capable, that the view which most naturalists until recently entertained, and which I formerly entertained – namely, that each species has been independently created is erroneous. I am fully convinced that species are not immutable; but that those belonging to what are called the same genera are lineal descendants of some other and generally extinct species in the same manner as the acknowledged varieties of any one species are the descendents of that species. Furthermore, I am convinced that Natural Selection has been the most important, but not the exclusive means of modification."

The Search for and Discovery of The Gene

The recognition that inherited characteristics flow through generations of both plants and animals existed in ancient times. A six thousand year old Babylonian tablet lists pedigrees of horses and possible inherited characteristics. Hippocrates, (460-375 BCE), postulated that each organ of the body grew from seeds that were passed from the father to the mother where they were reassembled. Aristotle (384-322 BCE), speculated that inherited characteristics were the result of purified blood shared during coitus, (we still talk of being related by blood.) Man was aware that an unknown mechanism was behind the transfer of inherited characteristics long before he became civilized, but had to wait for the advent of the scientific inquiry to discover the chemistry involved.

Gregor Mendel began to decipher the mechanisms responsible for inherited characteristics in the middle of the 19th century when he proposed that certain characteristics were carried forward to the next generation in discrete units. Mendel used a strict methodology. He observed and recorded results. He then hypothesized and followed up with further experiments and observations. His methods remain as the basis of genetic studies to this day.

The evolution of living things from disassociated groupings of single cells to multi-cellular life is documented in nature's hard cover books, (sedimentary rocks), and explained further in its soft cover books, (DNA). The examination of these records has given us a clear view of the developmental paths life has taken and of the twisted threads that make up nature's language of life.

Nature describes life one individual at a time using only a four-letter alphabet and simple grammatical rules. By allowing an occasional grammatical error and by demanding that variants comply with both internal functional requirements as well as changing environments, life persists. We call this thread language genetic information, each single fully formed description a genome, each set of threaded sentences a chromosome, and the discrete units described by Mendel, (sentences within chromosomes), genes.

The gene is the piece of the puzzle that Darwin needed to explain the evolution of life through natural selection. Now, we can see the gene, manipulate it, read its information, and trace the history of life recorded in its descriptive codes. The gene is now a tool of man as well as a tool of nature and has created a host of new questions.

As concepts regarding evolution coalesced, physical evidence was being accumulated. In 1869, a Swiss chemist, Johann Miescher, extracted a substance containing nitrogen and phosphorus from individual cell nuclei. We now know that what he extracted was the phosphate molecules that, along with sugar molecules, form the backbone of genetic strands and the nitrogen from molecules forming part of the nucleotides of DNA. These nucleotides were later determined to consist of four basic types, adenine (A), guanine (G), thymine (T), and cytosine (C) and to exist in both animals and plants.

Further examinations determined that every species contained different proportions of these nucleic acids and in the early 1950s, Austrian born biochemist Erwin Chargraff found that although proportions varied, the amount of (A) was always equal to the amount of (T) and the amount of (G) was always equal to the amount of (C).

At about the same time, using X-ray diffraction methods, British physicists Rosalind Franklin and Maurice Wilkins showed that DNA cast a shadow resembling a string of overlapping Xs. The shape of the shadow, and the relationship of (A) to (T) and (G) to (C) suggested to American geneticist James Watson and British biophysicist Francis Crick that DNA had the shape of a twisted ladder, (a double helix). In 1953, Watson and Crick constructed a large wire model of the molecule suggested by the information and it became clear that the couplings of (A-T) (T-A) and (G-C) (C-G) were rungs on the ladder.

Watson and Crick also noted that their model fulfilled the requirements for a hereditary molecule in that the rungs, AT and GC could be cut, leaving a single genetic letter attached to each side of the dissected ladder. They also noted that a new side could be replicated to replace the missing side by growing the appropriate missing pairs of molecular letters, resulting in two identical ladders. To confirm Watson and Crick's conclusions, geneticists Matthew Meselson and Franklin Stahl grew bacterial cells in the presence of heavy atoms of nitrogen so that both sides of their experimental hereditary ladders used only heavy nitrogen in the formation of the bacteria's nucleotides. They then placed the bacteria in a medium of normal nitrogen and as the bacteria reproduced, they used the normal nitrogen to form the missing sides of each ladder. As expected, the new bacteria contained equal amounts of heavy and normal nitrogen, supporting the hypothesized replication process.

It is interesting to note that Aristotle foreshadowed evolutionary thought when he wrote:

"...and in like manner, as to other body parts in which there appears to exist an adaptation to an end and all the parts of one whole happened as if they were made for the sake of something and have been preserved by having been appropriately constituted by an internal spontaneity and further that things not thus constituted have perished and will continue to perish..."

Philosophy, science and religion have all contributed to the development of genetic theory and continuing investigations have expanded into numerous sub areas of practical application and further research. Classical genetics now includes:

- microbial genetics,
- population genetics,
- cytogenetics,
- molecular genetics,
- genomics,
- human genetics,
- behavior genetics,
- epigenetics,
- applied genetics in medicine, agriculture and industry,

Genetics is now out of the box and cannot be put back without dismantling civilization itself. We now know that every living form contains its own unique coded description, that the language is the same for bacteria, plants, and animals, and that all life has a common origin. We also know that life's code is passed from generation to generation ready to be translated by RNA into proteins that make up all multi-cellular structures.

From atoms came the chemistry of life, from the chemistry of life came the coded word, from the coded word came living forms, and from the living form came *Awareness*.

Natural Selection

The Concept

The concept of "Natural Selection" is extraordinarily simple, almost an oxymoron, "what works continues and what doesn't work disappears". The concept of natural selection can also be applied to the study of inert physical formations and elements, but is only significant when applied to the development of living matter.

Darwin makes it clear that his observations taught him, not how things evolve, (he disliked the term, *evolution*, and preferred, *transmutation*), but that variations in living forms were naturally selected by being tested for their compatibility with their environment. In the introduction to "The Origin of Species", he states:

> "......I am convinced that Natural Selection has been the most important, but not the exclusive means of modification,"

Darwin understood the process that nature uses to weed out and promote living forms according to their compatibility with their environment. He was also aware that living things were constantly changing through the modification and abandonment of various characteristics. What Darwin could not explain was how these new attributes came about. For natural selection to work the reproductive process has to, occasionally, produce a new or altered characteristic. This continual process of modification was observable and accepted but was difficult to explain. The controversies that surrounded the term "Evolution" were primarily the result of a lack of information regarding the modification process. The "how" controversy continued for many years until Mendel's proposal that living characteristics were transferred in discrete units was re-discovered. Some argued that living variations were the direct result of a purposeful push by the environment. Others argued that the variations were caused by increased or decreased use of an appendage or organ, and others argued that they were unexplainable random occurrences.

Darwin's choice of words to explain nature's process of sorting through emerging attributes created controversy because he chose the word "selection", which carries within it the hint of a purposeful decision. This was not Darwin's intent. He fully understood that the process is mechanical, not purposeful. Darwin's insight into nature's method might have generated less controversy had he chosen, "natural elimination".

From this perspective, living attributes carried forward and incorporated in future generations are "left-over", after attributes that are detrimental or insignificant are eliminated. If Darwin had chosen to describe the process nature uses to test new life against the environment as, *Transmutation through Natural Elimination*, he may have been better understood, but probably would not have secured a place in history.

With the discovery of the gene, we now understand how living forms are randomly modified and how these random modifications are sifted by natural selection to add their survival value to future generations. Both steps in the process, random modification and selection by elimination are required for life's continued adaptations. Natural selection is a very simple process, a very efficient adaptive method, and a powerful force. The principles of natural selection explain more than just life's physical attributes. Natural selection has played a role in the advent and development of,

- sensory organs,
- neurological centers for interpretation and response,
- long and short-term memory,
- like-kind recognition,
- language and much more,

Natural selection has been at work selecting more than just physical attributes since life began. Natural selection has also been the impetus for the advent and advancement of *awareness*.

Applying the concept of natural selection to the evolution of awareness makes the state of awareness understandable. Attributes of awareness are tested against environmental conditions just as physical attributes are tested against the environment. When they enhance survival and or reproduction, they transfer to future generations by gene transfer, by example, and through language. If an attribute of awareness is detrimental, it is naturally 'deselected'.

Natural selection explains the physical diversity and history of life. It also explains the development and diversity of awareness. Natural selection is not only nature's guiding hand; it is also nature's alarm clock and has been busy awakening organic materials as living forms developed.

The Process

After a few billion years of natural selection, some generalizations regarding the process can be made. When exposed, these generalizations seem obvious because they are observable extensions of other familiar natural processes.

- Natural selection requires continual deviations from pure strains
- Natural selection filters toward complexity
- Natural selection filters toward accelerated selections
- Natural selection can be focused or expanded
- Natural selection filters toward advanced awareness

The selection of a new characteristic requires at least two different attributes from which to choose and a standard of measure against which they can be tested. Neither of these conditions are purposeful arrangements. Physical environments are random, endlessly diverse and ever changing. Add to this physical complexity, the living aspects of an environment, including predators, parasites, essential foods etc. and the complexity of the total environment against which our two choices are to be tested becomes extreme. Tested against such complexity the odds are; that if only two options are offered, both options will fail. Increase the number of attributes offered for testing from two to a very large number, and the odds of a successful selection are greatly improved. Natural selection functions effectively only when the diversity of the natural environment is challenged by an equally diverse array of fluctuating living attributes. A single strain in a static environment will never change and a pure unchanging strain challenged by a changing environment will not survive.

The simplest pieces of reproductive matter gave birth to the simplest of living cells. Anaerobic bacteria thrived for several billion years in earth's early hostile acidic, hot, methane environment as life's first and only kingdom.

As the earth cooled, cyanobacteria emerged and over several billion years, transformed the earth's atmosphere from 1% oxygen to 20% oxygen by excreting oxygen through a process known as photosynthesis.

The chemical transformation of a planet's entire atmosphere by microscopic life is not only amazing in its scope, but also in its results. Without the atmospheric transformation there would be no ozone layer, no life on land, and the other four living kingdoms could not have evolved. This first kingdom of early single celled life left records in sedimentary rocks and as fossilized stromatolites. Blue-green bacteria are alive today and, along with a few surviving anaerobic bacteria, are life's longest persisting forms.

Kingdom 1, Prokaryotes, (bacteria), still live around, on, and inside us, in extreme numbers. There are over 10,000,000,000 in every spoonful of garden soil and they comprise a significant portion of the dry weight of all animals. Over 10,000 species of bacteria have been identified and there are many more. Bacteria continue to play a significant role in Earth's biosphere, including the chemical composition of the atmosphere. They also provide essential symbiotic survival relationships for most plants and animals, including man, are both the cause of many diseases, the basis for many cures, and are themselves subject to viral infections. All bacteria belong to the kingdom prokaryotae (monera). Early bacteria lacked a well-formed cell nucleus and reproduce asexually by simple cell division, (mitosis), and in this sense were immortal.

Bacteria were the only life forms for more than half of the history of the Earth. They begin to leave a fossil record three billion years ago. With extreme numbers of bacteria tested by an ever-changing environment over such a long period, one would expect a proliferation of other types of living forms. Natural selection, however, continued to test only bacteria for two billion years, producing hundreds of identifiable species, and did not jump to more advanced forms of life until cyanobacteria transformed the atmosphere.

With the advent of a new oxygen rich environment, natural selection took advantage of the new energy source and cells with a nucleus quickly evolved in complexity and began to experiment with early sensory adaptations. From these eukaryotic cells, multi cellular life emerged with a level of awareness required for advances in mobility and life on land. Working in tandem, advances in physical attributes and sensory aware states, accelerated the process.

After a very long period of single cells as the only living form on Earth, kingdom 2, protoctistae emerged. These multi cellular forms now consist of,

- protozoa,
- aquatic kelp,
- slime molds
- and molds

The first multi-cellular forms began about three hundred and fifty million years ago. Protozoa, kelp and molds persist today with twenty-seven phyla and thousands of species. From kingdom two the other kingdoms developed.

- Kingdom 3 Fungi
- Kingdom 4 Animals
- and Kingdom 5 Plants

Each kingdom began their separate lines with the earliest evidence for aquatic animals appearing about seven hundred million years ago, and the earliest land plants and fungi appearing about four hundred and seventy million years ago.

Today nearly all animals are aquatic and worm like with only two of the thirty-three primary phyla truly adapted to living on the land, (chordates and arthropods). Thousands of species of animals exist today in these two phyla and tens of thousands of species have arisen in all of the animal phyla, with most now extinct. Plants have fewer phyla with only ten but have tens of thousands of species. Fungi are divided into only five phyla but also have thousands of different species. The large numbers of species in the five kingdoms attests to a natural divergence from pure genetic strains and the augmentation of diversity by the natural selection of variant forms. The increased complexity of later species attests to a natural trend toward complex forms including increases in awareness.

The selection of ever-increasing complexities associated with cilia, flagella, fins, legs and wings attests to natural selection's tendency to select the more complex to improve mobility and insure survival. Likewise, selections of complex manipulative apparatus like mandibles, beaks, tentacles and hands, helps to insure life's future through the advantage of natural tools. The interactions of predator and pray have also led to complexity in the natural selection of weapons like; fangs, claws, lures and poisons, and defenses like; body armor, natural tazers, camouflage and natural antidotes.

Complex sensory and neurological systems are naturally selected as a necessary accompaniment to increases in physical complexity. In combination, increasing complexity in form, paired with increases in awareness, has provided a survival advantage in rapidly changing environments and in competitions with other evolving forms. Natural Selection appears to have rejected the path of "keep it simple and stupid" in favor of selecting, "the complex and smart".

For the first two thirds of life's four billion year history, life evolved at a very slow pace. The natural selection process was in low gear because the environment was somewhat static. Only the oceans supported life and proto genetic forms had produced only single celled creatures. Natural selection also remained slow because there were few new attributes to be tested and environmental tests varied only slightly.

There are several theories as to what shifted natural selection into high gear.

1.   That the prodigious success of anaerobic single cells multiplying at an arithmetic rate breathed up most of the methane and carbon dioxide and exhaled so much oxygen that they poisoned themselves, created a new atmosphere, and opened the door to a new type of cell that used oxygen and exhaled carbon dioxide.

2.   That the entire Earth became frozen, that the ice absorbed most of the carbon dioxide in the atmosphere, and when the ice thawed, millions of years later because of volcanic activity, the carbon dioxide was kept from re-entering the

atmosphere by being absorbed in the melt water. With the ice and most of the carbon dioxide gone, the Earth presented a new and more complex environment that spawned new and faster developing life forms.

Whatever the cause, we verify the acceleration of natural selection by examining fossil records. Natural selection was slow when mitosis (simple cell division) was the only reproductive method and the biosphere was relatively stable. The advent of gene sharing introduced sex and death simultaneously and increased the number of variant genetic combinations exponentially. Instead of waiting for an anomaly to occur in the static genetic arrangements of single cells, gene sharing produced new genetic arrangements with every swap. Natural selection requires large numbers of potential selectees to find adaptations that improve survivability and pass on modified codes. Codes carried forward by variant forms carry within them the tendency for even more variant forms and accelerate their production. In this convoluted way, gene sharing, (sex), caused an explosion of variant forms to run through the natural selection process. New forms soon found suitable niches in nature's environmental offerings and continued their own genetic combinational experiments to create an even greater pool of variants. The proliferation of new forms also added to the complexity of the biosphere's testing ground and opened many new niches into which new variants could benefit from even smaller advantages.

New forms that generate many new forms survive and replicate more frequently because the environment becomes more diversified as they add possibilities. The result of many new combinational trials is a constant push on natural selection's accelerator.

Natural selection favors genomes with tens of thousands of genes and millions of alleles because they are prodigious variant producers. The preference by natural selection for large numbers of alleles in genetic structures, (the unit of transfer for physical characteristics), also applies to the selection of advanced aware states through an increase in meme choices, (the unit of transfer for acquired responses.) (see chapter five). The pace of natural selection accelerates with each success and this tendency contributes to both the accelerated development of awareness and physical forms.

CHAPTER TWO

Chapter Contents

CHAPTER TWO

AWARENESS DEFINED

Awareness is the essence of our being and is evident in all life, but like the air around us, awareness is invisible and, unless disturbed, goes unnoticed and unstudied. Lacking physical attributes, awareness, as an emergent state of creative energy, has eluded serious scientific inquiry. Recent investigations into sensory organs, the neurological centers that process sensory information and the evolution of these organs are beginning to expose awareness as a common aspect of life, and are broadening our perspectives.

Language is humanities greatest asset. It allows us to cooperate, create and explore, but it can also limit our ability to define and understand ephemeral phenomena. The emergence of life from inanimate matter and the awakening of matter in living form are one such phenomenon. Our vocabulary encompasses concepts like intelligence and cognizance but we have no word or concept to encompass the general awakened state of matter and must use our closest defining word, awareness, and try to stretch it to fit.

We are beginning to realize that awareness, as a larger concept, is essential for life's continuance, is at the center of our existence, and defines all of life's activities. Awareness is the reason the flower opens to greet the sun and the power behind human inventiveness. The essential attribute insures life's survival, is the capacity of living things to monitor and react to internal and external stimuli, and is the essential attribute that gives life a sense of self and the ability to identify like kind. Awareness is produced by and is dependent on, life's physical organs, and has no physical attributes, but it can be measured, studied, modified, augmented, created and destroyed. It has evolved in concert with life's physical forms, has an evolutionary history of its own and has been an equal partner to genetic evolution.

In man, awareness has gained a dominant position over genetics in natural selection, and if we are to understand evolution in full measured and maintain control of our destiny, we need to account for awareness as a state of awakened matter equal in importance to physical attributes.

Awareness As An Essential Element In Evolution

Elements of awareness have been noted and examined for centuries. Human behavior, animal behavior and the emotions of both man and animals fill the earliest of written records. Darwin began to categorize animal behavior as he referred to instinct, reflexes, and emotions as ways in which animals respond to events and situations. Other attempts to give animal and human awareness a base for scientific inquiry include the study of mimicry. Darwin devoted considerable time to giving examples of mimicry and to presenting the views of various naturalists and their interpretations and definitions. A concept for an ephemeral entity called a "mime" was introduced in the twentieth century as the behavioral counterpart of the physical "gene" claiming it was a major contributor to the evolution of behavior. As a deductive starting point, this concept held promise but lost momentum when popular writers applied the mime solely to human behavior patterns to promote utopian social orders. In spite of this early set back, the potential for the concept of mimicry to help explain natural selection in advanced life forms remains viable and is worthy of reexamination.

To appreciate awareness as distinct from biological processes requires a conceptual disconnect from the chemical/ mechanical view of science. Awareness is linked to the biological and genetic aspects of life but has its own attributes and contributes to survivability in ways that cannot be explained mechanically. The simple reactivity of single celled life has evolved to become the complex choices of animals and the contemplative moments of man. These advanced states of awareness cannot be explained by genetics or simple reactive responses. Choices made and contemplative conclusions reached within these states impact survival equally, or even more strongly, than genetically imparted autonomic responses. The complexity of cellular forms, advanced sensory organs, and the growth of neurological centers for the processing of sensory inputs, has produced a level of interaction between the living form and the environment that exceeds mechanical biological states.

Just as multi-cellular life exists at a level above, but not separate from single celled life, awareness exists at a level above, but not separate from, living biological forms. Without awareness, even in its most primitive state, life is not viable. The state of being alive requires more than just the abilities to collect and produce energy, separate itself from its surroundings with a membrane, and reproduce. Life requires the ability to interface effectively with its surroundings in order to collect the nutrients and energy needed for sustenance and this interface, in its simplest form, is the progenitor of awareness. Awareness at its most primitive level is only chemical mechanical reactivity. Add sunlight to a chloroplast and living energy is produced. Drift into an area of chemical nutrients and thrive, drift out and die. Develop the ability to survive periods without sunlight or nutrients by suspending certain internal processes, and survivability is improved. These reactions require the ability to sense changing conditions, by an internal or external monitor, and the ability to react. These simple reactive changes of state are primarily chemical/ mechanical responses, but they are also the first evidence of an ascending level of awareness. Each small improvement in a living form's ability to sense and react to changes in its environment improves its chances of survival and promotes the natural selection of advancements in aware states. The ability to sense and interpret, chemical surroundings, the intensity of light, vibrations, temperature, electrical fields, magnetic fields, motion, and one's internal conditions, all have the potential to improve survivability. Evolved organs for advanced awareness are the result of successive small improvements in awareness accumulating as survival advantages.

The Development of Organs of Awareness

The ability to react to external conditions creates a tremendous survival advantage. Driven by this advantage life's increasing complexity has produced an array of advanced sensors and complex neurological systems that are common to most living forms. Observe/react capabilities are grounded in cellular genetics, but when the process reaches a level of complexity where random reactivity is possible, abilities beyond simple physical reactions are created and a new arena for natural selection is opened. The activity of the simple protein flagella of prokaryotic bacteria and the more complex activity of the flagella of eukaryotic cells may indicate a sensory purpose as well as mobility. Flagella activity can be stimulated by an external prompt to swim upstream toward a source of sugar, and the motor activity may be a precursor to other more complex chemically reactive organs.

When a simple living form reacts to an external stimulus directly by using a genetically ingrained biomechanical mechanism, the sensory/ response connection is direct. In slightly more advanced life forms, with the ability to recognize nuances in stimuli and respond with several options, the response is not direct, and choice is introduced. Having the ability to be more discerning and to react selectively, through options, adds survival advantage. This new arena for natural selection I call; active awareness, (as opposed to passive awareness).

Like all of life's forms and processes, active awareness has developed slowly through the trial and error process of natural selection. Drawing a definitive line between simple reactivity and the complex reactivity of aware choice is difficult. Life developed slowly from natural chemical processes into replicating chemical units, into simple single cells, into more complex cells, into cell congregations, into complex cell congregations, and into multi cellular forms. Life developed in a continuum, and driven by survival advantages has naturally selected increased cellular complexity and increased awareness.

Awareness and physical complexity are co-joined and interdependent. Active awareness, (the ability to choose) has become a separate but equal player with the gene in determining life's future. Life's early physical aspects were connected to early awareness by precursor cells to sensory and neurological organs. From the simplest to the most complex arrangements this connection between life's physical and aware states offered up new possibilities for natural selection and gave favored status to the development of advanced sensory organs.

Being able to sense a variety of environmental conditions and respond appropriately improves a living form's chances to avoid elimination. Surface receptors in a single cell that are responsive to chemicals at high levels of concentration add survival value, but the ability to sense the same chemicals at much lower levels of concentration, or to discern changing levels of concentration, adds an even greater survival advantage. Natural selection has filtered out the less discerning and favored those capable of measuring and monitoring in greater detail, Multi cellular life, from the early layering of single celled families to the telescopic eyes of birds of pray, have had their organs of awareness naturally selected because of their advantage.

Without sensory organs, awareness is an empty state. As sensory organs evolved, they improved in sensitivity, and allowed life to examine its surroundings in greater detail. Each species collects and creates a different image of its surroundings using its own unique sensory and neurological systems. Each of these living images is but a glimpse into total reality limited by the abilities of sensory receptors and the capacity of neurological processing centers. The totality of all the images formed by all living forms in every environment creates an ever-expanding view as the universe awakens, but the view is fractured by life's individuality. Life senses its surroundings one individual at a time because gene swapping between individuals has proven to be nature's best method for accelerating natural selection. Natural selection shifted into high gear with the advent of meiosis, and is shifting into an even higher gear as language in advanced life forms, especially humans, creates group awareness and new survival advantages.

Sight began as a simple cellular sensitivity to light, hearing as a simple cellular sensitivity to vibration, smell and taste as simple reactivity to nearby chemicals, touch as a simple reactivity to pressure and temperature. Natural selection has brought us to our present advanced ability to sense our surroundings using complex organs of awareness because of their survival advantages. The survival value of advanced awareness is so influential in natural selection that one could speculate that the evolution of body parts for subsistence and mobility are subordinate to the evolution of awareness, and serve only to provide transportation and nourishment for sensory and neurological organs.

Several single celled species of phyla *dinoflagellata* have eyespots with a light sensitive layer of carotenoid pigments covered by a clear zone. One species, *Erythropsidinium pavillardii*, has a more complex oculus that it apparently uses to detect pray. The oculus of this species has a pigment cup covered by a fluid filled chamber and a lens that can change shape and the entire oculus can be protruded from the cell and point in different directions. *Dinoflagellata* are primitive single celled creatures that existed in the late Proterozoic without a well-defined modern cell nucleus and yet they developed rudimentary organs of sight from their chloroplasts. Even at this primitive stage in the development of living physical forms, natural selection was filtering into existence cell appendages, *undulipodium*, for mobility and *ocellii*, for vision. The early introduction of organs of awareness as an essential supplement to genetic selections for physical attributes is also evidenced by the early evolutionary partnership between the gene, (the unit of transfer for physical traits), and the meme, *(pronounced meem)*, the unit of transfer for acquired responses, *(more about memes later)*.

Sensory Methods to Awareness

Sensory awareness serves two purposes.

- Sensory capabilities that examine and monitor internal conditions
- Sensory capabilities that examine and monitor external conditions

Both of these monitoring systems utilize a variety of specialized cells and organs to gather information and both depend upon connections to various response and control centers that prompt reactive behaviors. In simple living things, these two sensory categories are primitive. A simple sensory trigger sends a signal through a primitive information pathway to prompt an autonomic survival response. These primitive states of awareness are observed in both plants and animals. Deciduous trees respond to a change in seasons by dropping their leaves, flowers open and close in response to daylight, bacteria become encysted in hostile environments, and slugs turn away from salt crystals. All of these are examples of simple sensory response mechanisms, (passive awareness).

Early response pathways are the precursors to more advanced neurological systems. In slightly more advanced life forms, the pathways to response centers have become more sophisticated to accommodate more developed sensory capabilities. Advances in sensory methods are accompanied by advances in neurological capacity. Advances in sensory capabilities have no advantage without a corresponding advance in interpretive and responsive abilities. Sensory and neurological systems are classified as separate systems physiologically but should be considered a single evolutionary system, a system providing survival advantages only if functioning together. Evolved sensory organs are life's windows on the universe and advanced neurological systems are the internal canvas upon which sensory pictures are painted. Awareness is the mirror in which the universe examines itself.

Natural Preferences in Sensory Development

The game of chance played by nature to create living forms has several options. The most successful and most used option is *natural selection*, but there are others including, *neutral selection* and *detrimental selection*. Body forms and physical characteristics have evolved into hundreds of thousands of adaptive living shapes, each a temporary best fit in a world of ever changing conditions. Some basic characteristics such as respiration and the ability to adsorb nutrients remain common, but in general, nature allows endless trials and tests and we are still discovering new species and ancient fossils with an endless variety of characteristics. The variety in living forms is the result of a completely random, endless, repetitive, and undirected trial and error process with over ninety percent of nature's trials ending in extinction, a dead end, but within this poor rate of success, a few characteristics persist as common advantages and carry through extinctions to become useful attributes for all forms. Sensory and neurologic capacities are two of these common advantages. Aside from their obvious survival value, these attributes persist because there are fewer options in sensory development than in body size, shape, or form, and therefore less chance for a matching error with the environment.

There are only a few useful ways to measure and monitor one's immediate surroundings or ones' internal conditions, and the restricted options have insured the natural selection of increasing awareness. Beginning as a pinhole view by an early life form viewing a centimeter of its surroundings, awareness has evolved to become mankind's enhanced representations of the entire universe.

Sensing objects and chemicals in contact with an outer membrane has developed in complexity to become sensory responsive organs capable of chemically analyzing proximate matter, pressure, and temperature. Sensing objects, radiation and motion at a distance also provides survival value and has prompted the evolution of organs sensitive to vibrations, (hearing), the ability to detect and analyze minute traces of chemicals dissolved in liquids and gasses, (taste and smell), the ability to feel radiated heat, (infrared sensitivity), and the ability to sense and interpret electromagnetic energy, (sight)

Vision has evolved to be the most prevalent and most important sensory method in creatures that live in the light. A dependency upon solar radiation for energy has developed in complexity to become sensory organs capable of interpreting electromagnetic energy in several forms, infrared, ultraviolet and visual wavelengths. These sensory sensitivities began very early in the evolutionary process and continued to be refined as they improved the survivability of mobile life forms. There is genetic evidence that single genes underlie all complex eye designs and all eye designs may be diversifications of a single prototype that first emerged in trilobites. There is also evidence that the evolution of the eye is not convergent but homologous. Eyes are present as eyespots in microscopic algae, as eyecups in flatworms, as compound eyes in dragonflies and other arthropods, and as retinal eyes in vertebrates. The diversity of eye designs is the result of their importance as a survival tool and of their natural selection in divergent forms for continued genetic inclusions. In embryos, the human eye develops outward from brain cells and the optic nerve cuts through the back of the retina creating a blind spot. In mollusks, the eye develops inward from skin cells with its nerve cells beneath light sensitive cells and has no blind spot. Proto eyes also exist in great diversity as the pinhole eyes of giant clams, the single chambered eyes of the nautilus and the compound eyes of certain clams.

The Development of Neurological Complexity

Naturally selected advances in sensory capabilities create a survival advantage by providing a wider and more detailed view of ones surroundings, but without neurological advances developing in concert, the survival advantage is lost. Sensory capacities and neurological capacities evolve as a unit, and their pairing in genetic selection is further coupled to the selection of physical abilities that use the increased state of awareness to an advantage.

Over time, the genes producing these mutually advantageous characteristics, sensory, neurological, and physical, become linked and a repeating genetic instruction. Being aware that danger or food is approaching by sensing vibrations in the water has no value if there is not an appropriate response, and appropriate responses require interpretive and triggering mechanisms to produce an appropriate physical action. The physical pathways along which sensory information travels from sensory organs to mechanisms for response have been genetically selected and, as recent studies have revealed, may have had their selection accelerated by a narrowing of options caused by restricted genetic pools and a rapid proliferation of normally sterile hybrids somehow passing on favorable genetic codes. The results of these accelerated selections, whatever their cause, enhance the process of sensory signals traveling through complex neurological control centers where they are compared to ingrained templates, a response chosen, and an activating signal transmitted triggering a survival response. We don't understand the full process in advanced life forms, but it is clear that better sensory information, coupled with more stored templates, allows nuanced choices that have a distinct survival advantage. Because of this advantage, the complexity and capabilities of, (observe, decide, react), systems have been selected in great numbers and have resulted in subtle and complex, neurological systems that are prevalent in all mobile life forms.

The ability to retain and repeat responses

Genetic drift and environmental changes create continuous variations in body arrangements and physical attributes. In contrast, neurological systems continue to evolve with similar characteristics. The evolution of awareness follows a much narrower path than physical forms. Sensory systems and neurological pathways are subject to genetic drift, but having a high survival value; detrimental changes are corrected and naturally selected, back, toward their optimum

One of the most amazing characteristics commonly shared in the diverse array of observant creatures is the ability to store and recall records of previous responses. Genetically ingrained bio/mechanical instinctual responses are without options. Stimulus response actions using these mechanical pathways are direct and unalterable. They fit the situation and succeed or do not fit and fail. Active awareness, on the other hand, goes beyond immediate reactivity with the ability to record previous neurologically responses and the ability to use these recorded events as templates for future situations. With the ability to store and recall information, memory became one of life's most valuable survival attributes and life moved to a new level of natural selection where awareness and the gene became full partners.

Mimicry

Mimicked behavior is observed in mammals, birds and many other species, but mimicry in its simplest form is difficult to distinguish from genetically transferred behavior. The dragonfly instantly masters flight after a metamorphosis and several years of living under water as a nymph. It searches with a compound eye for a mate in the air above its watery birthplace and exhibits skilled flying abilities and amazing threat avoidance without having had any opportunity to acquire these skills through mimicry.

Others on the evolutionary chain use mimicry directly as a tool to instruct their young in survival techniques. Instinct and mimicry work in tandem to create the behavior of most advanced life forms, including Man. To categorize these distinct methods of transferring survival information we attribute the mechanical transfer of survival information to genes and survival information transferred by mimicry to memes. Distinguishing the two in complex behavior is difficult because they often complement each other in the same behavior or response. The behavior of the dragonfly in flight may be evidence for genetically encoded awareness, a very complex form of adaptive memory templates, (genetic instruction), associated with species that undergo metamorphosis as a part of their life cycle. (A genetic capacity with potential worth investigating)

For life forms not blessed with a full set of survival performance options at birth, the ability to mimic behavior and learn from observation requires acute sensory ability and the neural capacity to retain and recall observations. The *'remember and repeat'* process also requires the ability to *'select'* an appropriate behavioral response. All of these abilities are the result of natural selection. They were not evolved to enable mimicry but have been blindly ingrained because memory and the ability to make appropriate recalls add survival advantages.

With the advent of active awareness, ingrained behavior patterns that took hundreds of generational genetic experiments to adapt was replaced by the ability to observe and choose based upon an evaluation of possible outcomes. With the advent of advanced awareness, the long mechanical process used by the gene to transfer behavior patterns was also suppressed as intelligent choices assured survival without the need for instinctual reactions.

The level of awareness needed for mimicry requires an advanced neural processing center, (brain), and advanced sensory capabilities. Mimicry developed slowly and in concert with complimentary physical attributes. Mimicry exists in only a few species. Its effect on those few species however, especially on Man, is profound.

The advantage of diversity in physical forms is, in some species, being overshadowed by the advantage of a joined aware state. In these species, group awareness is beginning to trump physical adaptation through natural selection making genes coding for advantageous physical attributes recessive, and genes coding for neural networks enhancing language and group awareness preferred. A cooperative communicative pack of wolves is more effective in bringing down large game than a single stronger predator, and humankind now rules the planet because of his advanced state of shared awareness.

CHAPTER THREE

Chapter Contents

CHAPTER THREE

AWARENESS AS A USEFUL STUDY

Beyond egocentricity

The scientific approach to understanding our surroundings and ourselves is very new. With a history of only four hundred years, the scientific method of observe, analyze, theorize, test and verify, has had spectacular success in discovering, and describing in detail, physical materials, living forms, and the rules that govern them. Human perspectives however, change slowly. We all tend to look in the same direction and use socially accepted standards to evaluate both our surroundings and ourselves. For most of human history, active observing gods were the explanation for most events and elaborate descriptive constructs were created to describe these unseen beings. With all questions answered by priests and theologians, few additional questions were asked and we continued to look in the same direction, believing in the same things, and using the same rhetoric.

(And then came Copernicus)

The idea that the sun and not the earth was at the center of the solar system was probably first posited by Aristarchus of

Samos around 340 BCE, but the idea remained dormant until the early sixteenth century when Nicolaus Copernicus, using Pythagorean calculations, authenticated the heliocentric concept. His mathematical proofs turned the earth centered Ptolemaic system on its head and forced the church to revisit its theological contention that the earth occupied a central position simply because of the importance of God's ultimate creation, 'Man'. Copernicus, unlike Galileo, was careful to avoid a direct confrontation with the church by dedicating his findings to the Pope

Awareness as a natural state, rather than as a spiritual or epistemological state, is not a new concept, but it is still not a favored perspective. Awareness as a natural attribute of all living forms is easily observable and we immediately identify with the animations of even the lowly earthworm as it expresses awareness when it encounters new surroundings. We have common ground with all living entities in the expression of awareness. Why then, do we continue to quarrel over its nature and continue to neglect any real investigatory efforts? Part of the reason may be the ephemeral nature of awareness. Unlike atoms, organs, and molecules it has no physical properties yet awareness can be measured, altered, and destroyed. Another reason for our avoidance may be an unwillingness to replace the idea of a soul with a naturally awakened state.

It was difficult to accept the sun as occupying a central position after generations of believing that center stage was reserved for the planet of Man. It is equally difficult to give up our special position as a separate creation of god and accept the idea that we are only the latest experiment by natural selection in advanced aware states. Another reason for the conundrum caused by the concept of awareness as a natural attribute may be the inherent difficulty of resolving our own self-awareness through self-analysis.

When we try to analyze self-awareness using our *self-awareness* we take a schizophrenic step that creates a hall of mirrors with our own reflections extending to infinity.

Many great thinkers have struggled with this problem.

St. Augustine of Hippo: "Rational thought is the servant of faith: 'unless thou believe thou shalt not understand.' *Isaiah.*

Rene Descartes: "cogito ergo sum." I prove my existence to myself by my awareness of myself but all else gained through the senses is fallible. Mind and body are two separate things

Nicolas Malebranche: "Whenever we think we are doing something, God is really doing it for us."

Benedict de Spinoza: "There is only one substance, and that substance we can conceive of as either Nature or God." Mind and body are part of the same reality, self-causing and self-contained.

John Locke: "The mind at birth is like a blank slate, waiting to be written on by the world of experience."

David Hume: "Knowledge can be gained only through the senses but causal reality cannot be inferred from sequence. Self is an illusion, nothing more than the continuous succession of perceptual experience."

George Berkeley: "esse est percipi" "To be is to be perceived. We can only know the ideas that are created in our minds as we observe an act or object. There is no way to prove that the object resembles our idea or that it even exists."

Ernst Mach: "we know only one source which directly reveals scientific facts – our senses."

What should be evident from the above and from other philosophical inquiries made during the age of reason, is the avoidance of any generalized inquiry into awareness as common to all living things. Equally evident should be our preoccupation with human awareness as distinct and separate from awareness in other life forms. The broader view of awareness as a shared quality of all living things continues to be suppressed by the egocentric obsession that we can best understand nature by studying ourselves. Awareness occupies such a central place in nature that it seems incomprehensible that we have not given it greater attention. The body of awareness around us is truly our universal home.

Establishing standards

We are programmed by evolved synaptic wiring to create informational nets that capture and hold observations for later categorization and use. A general study of awareness requires the same approach. Informational nets, as well as the placement of additional observations, are adjustable. What is important, initially, is not the order of groupings or information, but the perspective that awareness is a natural and primary attribute of all living things and that its attributes have been naturally selected just as physical form and function have been naturally selected.

Creationist theories need to be held at bay if we are to fully understand awareness as an essential element of life. Imposing imagined purpose into the study of awareness creates detours that end at stop signs with the word "God" imprinted on them. The study of awareness will inevitably enhance human relevance, not diminish it, and efforts to derail further inquiry will only delay the discovery of our true place in nature.

In the hope that a concerted effort in studying awareness as an evolved, central and necessary attribute of life can be initiated, I propose the following categories and attributes as starting points for an organized classification of awareness in all living forms. By establishing categories and attributes that can be identified and measured, awareness can be separated from anatomical and neurological investigations and from intelligence tests that prevent the viewing of awareness as a real attribute of living things that transcends its biological base. Examinations of specific points within various neurological centers and their connection to specific aspects of awareness help us trace the paths involved in the cooperative natural selection of advances in awareness and its biological base, but do not address the commonality that has been naturally selected in many different species and along many different paths.

Categories and Attributes of Awareness

Categories:	Attributes:
Primitive awareness	Pace of awareness
Passive awareness	Scope of awareness
Active awareness	Duration of awareness
Cognitive awareness	Retention and Recall
Social awareness	Malleability
Self awareness	Language

Possible areas for qualitative analysis

Mathematics has played a crucial role in the acceptance and substantiation of nearly every major scientific concept. Often, the addition of mathematical descriptions of observations adds credibility and allows new perspectives to stand against the resistance of tradition. The mathematical analysis of Gregor Mandel's observations of inherited characteristics led directly to the conclusion that information could be passed in discrete units from generation to generation, and that the discrete unit theory could explain the variance in forms noted by Darwin. The use of mathematical formulae by Kepler to revive and explain the idea that the sun and not the earth was at the center of the solar system added credibility and held the church at bay (temporarily). Later observations and records of celestial movements by a church official gave Galileo the information he needed to apply the ellipse to the movement the planets. Kant, a devout Christian, used the calculus to explain movement and gravity, and Einstein turned his mind exercises into tensor formulae to explain relativity. If awareness is to be accepted as a legitimate field for scientific study, it too must offer up specifics for mathematical analysis.

Before mathematics can be applied to aware states, a database is needed. The mathematical application that moved the idea of evolution from theory to an area worthy of research was made possible by a database built by a Monk obsessed with pea plants. Any expectations for a monastic data collection of aware characteristics without a religious taint however, seems unlikely. Observing and recording the many ways matter, in organic form, senses and uses, information from its surroundings is a study of ephemerals. To be relevant any study of awareness must focus on the broad spectrum of aware states beyond Mans' awareness, a difficult task when observers using there own aware states are asked to analyze other aware states. It will be a difficult task, but may be possible if egocentricity is kept at bay.

Divisions and comparisons using the categories of awareness described in the next chapter can provide a starting point. Levels of awareness grouped by similar characteristics may not match groupings based on physical attributes, lineage, or species. This divergence highlights both the interdependence of evolving physical and aware states and the differing environmental restraints that allow physical forms greater latitude. Even our most basic division between plants and animals will exhibit overlaps in aware states and as these differences are noted, a database will begin to form. As difficult as it is to link our synaptic thought processes to the holistic processes of nature, we are making progress.

It is the goal of this book to expose awareness as a useful general concept, to create a perspective that living entities are naturally emergent forms of organized and energized matter, and to expose emergent awareness as an extension of living energy.

CHAPTER FOUR

Chapter Contents

CHAPTER FOUR

CATEGORIES OF AWARENESS

Levels of awareness are not distinctly separated in nature and establishing categories for various levels of awareness is difficult,. Observed patterns however lead to general categories that can be useful for understanding awareness as an evolved state and for further research.

Primitive awareness

Primitive awareness; Basic chemical / physical interactions between organic and inorganic materials

The earliest precursors to living forms were little more than a few active chemicals trapped inside lipid sacs, a bubble of froth indistinguishable from the froth of dissolved chemicals surrounding them, but these sacs and the elements inside them were isolated from the undirected mixing outside and free to join and experiment in new ways. From the numerous experiments tried inside these tiny laboratories, a few resulted in a self-sustaining slow chemical reaction, and the first spark of life was ignited. Replication by these self-sustaining lipid sacs most likely occurred serendipitously in areas where essential chemicals were concentrated along with an additional element that caused the bubbles to break up into smaller sacs spreading the self-sustaining spark of life throughout the colony.

Awareness in these first replicating energy cells was only a simple chemical and mechanical reactivity to environmental conditions, but it was significant. It was the first case of a completely new arrangement of matter passing its first test of survivability. The new arrangement led to new forms and new tests, altered the environment, and gave direction to life's future. Life continues, in part, because of the ability of these first cellular forms to react to chemical and energy conditions, and their simple reactions are the root of awareness,

Passive awareness

Passive awareness; Stimulus / reaction events that occur and are repeated using simple sensors and primitive biochemical pathways

Very early in the development of life, nature added the advantage of simple biomechanical responses to its physical forms as a genetically advantageous option. Improvements in life's ability to sense and react to changing conditions supplemented much slower genetic adjustments and awareness became a valuable partner in adaptive survival. Primitive sensory organs existed early in life's development but passive response mechanisms remained life's only adaptive means for the first two thirds of its history. For several billion years life consisted of asexual immortal single cells reproducing by simple division and genetic adjustments were slow. Cellular complexity increased during this long incubation period and produced a basic awareness with enough value to evolve and improve in concert with genetic advancements.

Awareness in life's early living soup was primitive but not developmentally arrested. Over time, the advantages of single cells to interface selectively with their environment became preferred traits. Cells that could adjust slightly to the intensity of light, changes in salinity, occasional exposure to dry conditions or the ability to assimilate nutrients more efficiently, became the first forms with early passive awareness. As these adaptive abilities exhibited themselves in greater numbers, natural selection created a preference for groupings of cells with differing but mutually supportive characteristics.

Layered families of single cells supporting each other symbiotically improved their chances for survival. Ancient fossil hummocks provide evidence for these earliest precursors to multi cellular forms. Prompted by the survival advantages of cooperative cells living in close proximity, an exchange of genetic material between cells became inevitable and the advantages of early hybridization at the cellular level in these hummocks accelerated the process of adaptive natural selection, and passive awareness became an integral part of evolution.

Passive awareness continues to be the only level of awareness in four of life's five kingdoms; prokaryotes, protoctistae, fungi, and plants, and continues to persists in the animal kingdom as reactive responses for non-specific monitoring and emotional and reflexive reactions. In plants, passive awareness is evidenced by reactions to light, to the seasons, to water and soil conditions and by some predatory plants hinting at an active aware reflex in reactions to insect traps triggers. A clear distinction between any of our self-imposed levels of awareness is difficult. In animals, the distinction is difficult because a sensory signal can simultaneously prompt both a passive and an active response. The advantages of passive levels of awareness to early single and multi cellular life has made awareness a permanent part of life's development and an integral part of the Earth's biosphere.

Reflex and instinct

Reflex and instinct; genetically transferred neurological programs for responses and behavior.

Between passive and active awareness is reflex, (knee jerk reactions to a stimulus), and instinct, (pre-programmed complex behavior reactions). Both require a certain level of awareness to trigger a pre-programmed reaction but appear to use different genetically imparted pathways and response mechanisms.

Responding to touch, sudden sounds or suspicious movement, reflex, (a form of passive awareness), developed as an instantaneous response mechanism to capture pray or to avoid a physical threat. There is evidence for reflexive responses in very primitive life and these responses have continued to be naturally selected in advanced life forms. In advanced living forms, the genetic coding for these responses creates direct synaptic connections from primary sensory organs to primitive parts of the central nervous system and an immediate interface to motor and glandular control centers. In primitive living forms, genetic coding for these responses creates a chemical or mechanical tension that is released by a trigger mechanism.

Instinct is more complex, involves a great deal more pre-programming and interfaces with both passive and active centers of awareness. The units of transfer for these pre-programming and complex interface arrangements are not clear but appear to be genetic. The codes for instinctive behavior are included in many life forms and program for activities ranging from the ability to fly, mating rituals, migrations, and many more. Instinctive activities are often coupled with active awareness, allowing ingrained responses the ability to adjust to immediate variables. (An unusual genetic / cognitive connection)

Reflex and instinct both present evidence of a strong link between the unit of transfer for life's physical aspects and the unit of transfer for acquired information. The challenge is to understand how the gene, as the language for physical replication, can also code for behavior. Living forms incorporate naturally selected neurological pathways and naturally selected interfaces between sensory organs and motor responses because they help to insure survival in active and changing environments, but natural selection does not filter for actual body parts, nerves or brains. Natural selection only filters for the genes that code for advantageous body parts or advantageous aspects of awareness. This filtering occurs when genetic instructions for body parts or brains, not compatible with the environment are erased by eliminating the host carrying the instructions. Living forms are merely test vehicles for genetic information and the success or failure of these test vehicles determines which genetic streams are carried forward. Battles for the survival of the fittest are fought between random genetic arrangements using the experimental living forms they produce as soldiers.

Awareness and behavior are intrinsically linked, and both depend upon sustenance gathering host forms. If a gene codes for a long neuron with a faster input to a motor response, and the living forms incorporating the gene escape predators more often than those without the gene and its long neuron, the gene is carried forward and carries with it a change in behavior. If a gene codes for a larger central neurological center and with it an increase in reactive options that improves the survivability of the gene's host body, the gene is carried forward and carries with it an increase in neurons producing awareness. These adaptive improvements are not planned and do not occur because they carry, within them, a foreshadowing of an improvement. They are completely random and gain temporary status for inclusion in a genome only because they continue to pass successfully through nature's survival filters. The illusion that these slow changes seem directed and logical, results from our own genetic programming by the same random selective method. Logic is only a reflection of selected survival responses.

Instinctive behavior is the result of the same pre-programming by random genetic selection but involves a more complex neurological interface and the inclusion of active assessments of immediate conditions. Natural selection is self-accelerating. Random gene selections due to occasional AGCT misspellings increase selection options that are tested by environmental filters. The result of more options offered for testing is a faster pace for the filtering process and a more rapid elimination of unsuccessful genes, a more rapid accumulation of successful genes, and a more rapid distribution of favorable traits. Advanced aware states are the result of this cumulative and accelerating effect.

Instinct is the successful mix of reflex and aware responses. It is repetitive because of its hard-wired basics and is successful because of its connection to active aware states. The coding for this mix is complex and goes beyond our basic genetic perspectives. The discrete units responsible for the transfer of these behavior patterns may be incorporated with other genetic exchanges, but deserve recognition as a separate class. I propose that these complex units of instinctive behavior be called; *Gemes*, (*pronounced geems*), a union of genetic learning transfer and memetic information transfer. (See chapter 6)

Active Awareness

Active awareness: stimulus / reaction events that utilize developed sensory organs and complex neurological pathways.

Primitive and passive aware states define most of life's kingdoms and phyla. Passive aware states dominate four of life's kingdoms with aware states occurring almost exclusively in the animal kingdom. Most animal species, (thirty-one phyla), are aquatic with land animals making up only two phyla, (chordates and arthropods). Active awareness is evident in both marine and land animals, and is most likely a product of mobility. There are thousands of extant animal species and tens of thousands of extinct species. The numbers are large and make clear the large percentage of environmental compatibility experiments that have failed and the much smaller percentage that have succeeded and become the mobile aware creatures around us.

Establishing a distinct line between complex levels of passive awareness and the low levels of active awareness is unnecessary.

Some life forms will inevitably straddle our categorical divisions and fit into more than one category. Our categorizations for awareness are of necessity artificial but remain essential for discussion and further investigation.

Active awareness developed as an essential survival tool when life became mobile. Living forms drifting aimlessly or rooted in-place wait passively for environmental activity to provide nutrients or fertilization and need only a limited repertoire of responsive capabilities. Simple sensory capabilities sufficed to meet survival and reproductive needs until life begin to move from place to place, a few centimeters in the case of amoeba, or thousands of miles in the case of migrating fish or birds. Self directed mobility requires advanced sensory and responsive abilities and active levels of awareness became essential when life began to explore new environments. The survival advantages of advanced states of awareness quickly became essential genetic traits and made active awareness a permanent part of mobile life.

There are many ways for living forms to sense and evaluate their environment but only three primary sensory techniques have been naturally selected more often than others. The advantages of being able to hear, see, and smell drastically improved survival and procreation rates and have become more sensitive over time. Increased sensitivities create the ability to detect and differentiate between nuanced conditions, and the ability to detect conditions at ever-increasing distances.

Sensitivity to touch has evolved from the sensitive membranes of bacteria into specialized receptors that can sense; heat, pressure, pain and vibrations in a variety of amplitudes and intensities and in a variety of mediums. The ability to detect vibrations from distant disturbances in solids, air, and water provides early alerts to predators, mates and pray. Improvements in the ability to sense and assess faint vibrations developed quickly and hearing became a naturally selected attribute in most mobile forms living on land. Species as diverse as frogs, birds and humans utilize similar sensory organs for hearing, small fluid filled sacks, called cochlea, lined with tiny hairs that are sensitive to extremely small displacements. These common organs are able to sense a variety of external vibrations and translate them into neural signals that are interpreted by a corresponding part of the creature's brain. Interestingly, nearly all creatures capable of hearing, except mammals, can regenerate the tiny hairs essential to hearing. In mammals hearing loss due to these hairs being damaged is usually permanent. Hearing, as a primary sense and has provided the bridge to new levels of natural selection with the development of auditory languages. Hearing remains the least studied of life's primary sensory capabilities in spite of the primary role it has played in directing the natural selection of advanced cognitive abilities.

Communication between creatures of the same species takes many forms, visual, chemical and vibrational. The survival advantage of communications between like kinds is evident in both land and marine animals. Marine animals, other than marine mammals, prefer visual signaling to communicate while land animals and marine mammals prefer vibrational languages. In man, vibrational exchanges, (passing information through air), became the precursor to a symbolized visual form of vibrations, (writing), capable of transcending time.

Long before man invented machines to record the sounds made by the vibrating molecules of air, he scratched symbols on stones representing objects for which he also had a specific verbal utterance. Over time, the scratched representations became synonymous with the sounded names and the stones could be made to speak. Symbolized sound is what distinguishes man from the rest of life on earth and is what has created and sustains civilization.

Sensitivity to changes in the chemistry of ones immediate surroundings is a third sense used by both land and marine animals, and to a lesser degree in all life's kingdoms and has developed into specialized organs capable of detecting just a few molecules of complex compounds drifting in air or dissolved in water.

The ability to detect the presence of a mate, a food source or danger by smell or its related sense, taste, has significant survival value. Chemical awareness has remained tied to early informational pathways even in advanced life forms, and results in emotional and reflexive responses more often than other sensory signals. Smell and taste are closely related, and developed from the same early passive sensory ability of microbes to detect chemical changes in close proximity.

Sensitivity to various wavelengths of radiation is the third primary sense. Cells sensitive to electro magnetic emissions have developed into specialized organs capable of imaging a surrounding environment in detail through receptors sensitive to radiated and reflected electromagnetic energy. Sight is probably the most advanced and the most often selected sensory capability for all advanced life forms and provides the greatest immediate survival advantage. Advances in neurological complexity are closely associated with advances in sight. Of the three primary means of acquiring environmental information at a distance, sight is the most valuable.

Living forms view and interpret the world around them as neurological activity created by signals transmitted from their sensory organs. The images are filtered bits of reality restricted by sensory limitations and the limits of neurological capacities. To the observing life form, the internal images appear complete but they are far from complete. The internal images filtered through sensory organs are only narrow segments of the total spectrum. There is a full range of electromagnetic radiation to which life is blind, a cacophony of sounds to which it is deaf, and millions of chemicals and compounds beyond sensory abilities to detect. The natural selection of sensory capabilities has remained narrow accessing only those most important for survival. The cerebral capacities of any life form, including man, would be overwhelmed if more information from the environment were provided.

Active awareness has opened windows on reality that allow a sense of immediate surroundings far beyond the capabilities of passive awareness. The natural selection of advanced states of active awareness has also taken life to a level where language has become a preferred selection. Language, in turn, has prompted the development of cognitive awareness as the next major step in the evolution of awareness.

Cognitive awareness

Cognitive awareness; Stimulus / reaction events utilizing extremely complex neurological pathways to interpret and analyze situational conditions before choosing a response

Cognitive awareness is the result of genetic codes favoring neurological and sensory development, but genetics is not the sole impetus for advanced states of awareness. When awareness in a species advances beyond simple reactivity to include considered evaluations, choices begin to assist natural selection in determining which genes will be eliminated and which will survive.

Cognitive awareness is not exclusive to man. It exists in many species. It exists in the hunting and defensive activities of marine animals, the nest building and use of tools by birds, the considered choices of mammals, and the inventiveness of man. All of these activities attest to the survival value of an advanced state of awareness capable of considered and creative responses.

Cognitive awareness is evidenced when a living form recognizes a threat, a mating opportunity, or a social situation, and is able to evaluate the situation and choose an appropriate response. Cognitive awareness is also evidenced when immediate situations present a mix of elements that require sorting to select those that are relevant and when a complex event requires the ability to choose both an appropriate response and an appropriate level of response. This type of behavior exists in other species, but is observed most often in mammals.

Certain alligators have developed an advanced cerebral cortex and are capable of understanding and responding to human commands at a level similar to domesticated dogs. The common crow can recognize individual human faces and uses a complex language that may have been carried forward from its dinosaur ancestors. Many deep-sea creatures use an elaborate language of bioluminescence, (the most common mode of communications on earth), that rivals the nuance of mammalian vibrational languages

Some marine arthropods and some land insects think as a group using patterns of communication that mimic closely the neurological patterns of mammalian brains and exhibit group behavior that may be early levels of cognitive awareness.

Recognizing cognitive awareness in other creatures is difficult if we think of animals, capable of existing outside civilization, as wild and therefore a separate class, subordinate and inferior. To get beyond this egoistic perspective we only need to recognizing levels of awareness that allow animals to deal with complex natural situations outside our structured human environments. Situations that would challenge, or overwhelm, our high level of awareness if placed in a natural setting, are dealt with routinely by other animals. We have a great deal in common with our cognitive aware cousins and enjoy our special position on the evolutionary tree primarily because of several fortuitous genetic acquisitions and one key memetic acquisition.

From our genetic history, we have acquired upright walking, a large brain, an opposable thumb and an articulate tongue. The memetic acquisition that sets us apart is a written language. Without these attributes, we would be one of the wild classes of animals we degrade. With these attributes, we have created the cruel and creative organization we call civilization, have built great works, wondrous tools, and have explored our surroundings in detail. Our awareness has been expanded to an extent where our power of choice has usurped natural selection and now determines the future of all life on earth. Recognizing our advanced awareness as the product of matter becoming alive and as the new determinant of life's future after three billion years of testing, creates responsibilities as caretakers for an entire planet that exceed divine directives.

Social awareness

Social awareness; a collective aware state of shared information by a cooperating group of individuals

Social awareness is evident in most communal creatures but is even more ephemeral than individual awareness and requires a separate category to appreciate its impact on individual and group genetic selections and survival. In its simplest form, social awareness can be a symbiotic relationship or the ingrained behavior of an insect colony. In forms that are more complex it can be common adaptive behavior within a herd surviving through many generations, or swarm intelligence surviving as demonstrated by insects, birds, certain crustaceans, and jellyfish.

In man, the symbolic accumulation of shared knowledge has become the defining characteristic of his species and the determinant of his future. In man, social awareness has overridden both genetics and individual memetics in natural selection. Mankind's ability to control and manage the future of his socially aware state appears, so far, to be somewhat beyond his individual and organizational capabilities, and by default, leaves his future not only in doubt, but subject to a new level of natural selection, *new even for nature.* (See chapter 9)

Self-awareness

Self-awareness; The ability to recognize ones own awareness as separate from the awareness of others.

The beginnings of self-awareness are evident in many other life forms but only man has become consumed with his self-image. Self-awareness in man is a special form of cognitive awareness that creates an illogical loop whenever we try to analyze and define ourselves. Our hard-wired synaptic thought processes have been genetically and memetically selected to deal directly with the reality around us in conceptual packets. We facilitate the process by adding symbolic enhancements that allows us to use our neurological capacities more efficiently when dealing with natural phenomena. In the process, we have created symbolic representations of ourselves and our self-symbolism has created a cognitive gap between our actual self and our symbolic self that has curious results. See chapter 10.

CHAPTER FIVE

Chapter Contents

CHAPTER FIVE

ATTRIBUTES OF AWARENESS

Establishing measurable attributes for a non-physical state is difficult. Categorizing various levels of awareness as, primitive, passive, active and cognizant, with group and self-awareness as additional levels is arbitrary but follows our ingrained intellectual need to identify, sort, and order information. Our categorization of living things, using physical characteristics, into kingdoms, phyla, classes, orders, families, genera and species are equally arbitrary but serve as an effective simplification tool for further comparative investigations. Advances in awareness follow the growth of complexity in life's physical development but are restricted to a much narrower set of options. Unlike the multitude of complex and changing conditions available to shape the physical attributes of life, the radiation, vibrations, pressures and chemicals suitable for analysis by organs of awareness is limited. As a result, millions of diverse living forms use similar methods to analyze their surroundings. Awareness exists as a special attribute of life's physical state and has its own evolutionary history, its own record of natural selection, and demands its own categories if we are to study it effectively. Understanding awareness as an essential universal characteristic of all living things should help us appreciate our relationship to all life and the responsibilities that go with our own advanced ability to sense and understand the world around us.

Appreciating awareness as a part of life and the environment is important, but the emergence of observers formed from energy and inanimate matter may affect basic arrangements in the universe in other ways. Strange things happen when the universe wakes up.

Time as a consideration

Time is created when a physical object moves or there is a change in condition. In a universe where nothing moves time is static, (does not exist). Conversely, in a universe without even the smallest interval of time available, nothing can move and space is static. From this perspective, time and changes in position, or condition are synonymous. Reduce one side of the equation to zero and the other side becomes zero. To complicate the concept further, Einstein proposed that when things do move, time is relative. In effect, each moving object creates its own time relative to all other objects and together the total of all movements creates a universal time. To make time even more mysterious, Einstein declared that universal time imposes a speed limit on all individual motions. No object can move relative to any other object faster than the speed of light, which seems to indicate that time and the speed of light are somehow connected, (which makes sense because speed, by definition, is distance traveled during an interval of time), but how is the universal speed limit enforced? Why can't we simply keep adding intervals of speed and get ahead of light? The answer is time. The universal speed limit is enforced by time. Speed up and time goes slower, reach the speed limit, which would require using all the energy in the universe, and time stops. Stop everything from moving and time also stops but, if no one is watching, who cares?

Does time require an aware observer? Without an observer imposing a pace of awareness on the movements around them, time is like vibrations in the air when no one is there to hear them. The movements exist but have relevance only to other inanimate objects and other movements. Time, as a concept, is irrelevant until observed just as sound waves are silent until heard. Is this just semantics or does the awareness of living matter add a new dimension to the universe, a dimension without physical attributes but, none the less, one with relevance? One type of reality existed before the advent of awareness and a new type of reality was created when awareness was added

When the two types, (one a mix of energy and matter), and, (the other the fractured partial impressions of matter and movement held by aware living forms), are juxtaposed, reality expands. Awareness and physical objects influence each other when they connect. An observer attempting to determine if light is a wave or a particle exposes the connection. Whenever this experiment is conducted, the experimenter always finds what he expects to find because his awareness alters the results. Time is arbitrary until awareness is interjected, then it becomes measured. Measured time requires an aware observer to record passing events using its own pace of awareness. Time, as we know it, is the result of our observational rate. When we are unconscious, time, for us, ceases and we require some other measure to fill in the missing period. Time without any aware observers seems meaningless and our analysis seems to hinge on semantics until we expand our awareness by projecting ourselves beyond the earth at speeds which genetic evolution could not anticipate. Only then does our time and universal time become relevant. Quantum mechanics adds credibility to the argument for awareness as an interactive part in the mix of matter and energy by demonstrating that time, position, and outcomes, are altered by being observed.

Pace of awareness

The circadian rhythm for all life on Earth has been naturally selected by the spinning rate of our planet, by the repetitive seasonal patterns resulting from the 23 ½ degree tilt of our planet's axis, and by the ocean tides as they react to the moons orbit around the earth. No other planet in our solar system has rotational intervals even close to those of Earth. Each is unique and each is the result of natural causes. We expect the sun to set, winter to approach and the tides to rise and fall at certain intervals because of the motions of the planet upon which we have evolved. The general pace of living awareness on the earth has been predetermined, but there are variables in the general pace that have been shaped by other factors including natural selection.

Differences in the pace of awareness between species exist. We observe these differences as varying rates of mobility and response. A garden slug has a slow pace of awareness that is reflected in its slow physical pace across the garden and in its slow response to stimuli. A common housefly has a much faster pace of awareness that is reflected in its ability to move quickly and make very rapid response decisions.

Establishing an accurate numerical measure for the attribute of pace is probably not possible, but a sliding comparative gradient, based upon a selected norm is feasible. Any gradient used to compare a pace of awareness will become arbitrary when applied to subtleties as small as differences between individuals of the same species, but may be effective in comparing and cataloging gross differences between species. Where survival or reproduction require a faster pace we should expect to find naturally selected rates of awareness above those found in more secure environments. The tree sloth moves in slow motion not because it is lazy but because the natural selection of a faster pace was unnecessary and its energy levels are low.

Scope of awareness

Until Man began using tools to enhance his observational capabilities, the scope of awareness of living things, including man, was determined solely by the acuity of their sensory organs and the interpretive capacity of their neurological processing centers. The scope of awareness of early bacteria extended only a few molecular diameters from its cell wall as it sensed the chemical conditions in its surroundings or the presence of an adjacent object. The scope of awareness of a chloroplast may appear to extend further, in that it reacts to light that has travelled 93 million miles from the sun, but it only senses the light in a passive sense and gathers no details other than the fact that light is present. Life accesses its environment through a limited number of similar organic windows because the methods were naturally selected early as the most effective in providing survival value. Similarities between the observational methods of very diverse living forms are also due to the limited number of physical possibilities offered by the natural world. Radiation, electro-magnetism, vibration, pressure, and chemicals are the primary aspects of the physical environment life uses to analyze its surroundings. As life began to access information from these five environmental attributes, living forms developed specialized organs to explore and occupy new environmental niches and extended the scope of their sensory capabilities to a naturally selected optimum.

At first glance, it would appear that physical size would be a primary determinant in the scope of awareness, but mobility adds an additional element that makes the mix more complex. If we define scope of awareness as a combination of observed area when motionless, expanded observable areas due to mobility, and the ability to acquire detail, we can begin to make comparisons and establish a sliding comparative scale for various living forms. Insects are small and usually travel short distances during a short life span, but the exceptions are extreme, The Monarch butterfly migrates a thousand miles but appears to access only a limited amount of environmental information to make the trip. The Robin also migrates thousands of miles but appears to access a large quantity of detailed information from its environment including the use of its right eye to sense, (see), the earth's magnetic field. The Robin's scope of awareness and pace of awareness are above the Monarch on our sliding scale but below Man's.

Man has moved beyond all comparative scales by inventing and utilizing devices to enhance both his awareness and his mobility. In this respect, he has progressed far beyond any other known life form. We still utilize our basic naturally selected sensory and neurological capacities but have the advantage of a unique naturally selected meme, (a complex language), allowing our inventiveness to evolve and through which our sensory extension tools have been created.

Duration of awareness

Awareness is a natural attribute of life has spanned billions of years and shares its development with the evolution of living shapes. In its most basic state, awareness accompanied prokaryotic cells in there transformation to eukaryotic cells billions of years ago, and has evolved in both complexity and capacity in concert with living forms ever since. Viewed as a whole, awareness spans the entire history of life on earth. Viewed as the direct product of an individual living cell, awareness in individual spans only the life of the individual cell. Viewed as the awareness of a community of cells, awareness extends to the life of the multi-celled community. Semantics will always play a role in attempting an accurate definition of durations occupied by aware states, but for comparative purposes, duration of awareness makes sense only when defined as the duration of an individual living form's passive, active or cognitive awareness.

Awareness, as the activity of a living form to sense and interact with its environment is intrinsically tied to the continuance of the living form producing it. When the organism, or the organs, producing awareness cease to function, awareness ends. Instructing ones young does not carry the instructor's awareness forward, only the transferred instructional elements persists and can be transferred to subsequent generations just as the gene persists by being passed on to progeny. In the broadest sense, awareness spans the entire history of life. In the narrowest sense, awareness spans only the immediate moments of a living forms existence. In the most practical sense, awareness exists during all of the functional moments of an individual life.

For some insects, their span of awareness persists only long enough for the earth to turn less than once upon its axis. For some mammals and reptiles, their span of awareness persists long enough for the earth to turn upon its axis more than 30,000 times.

Retention and recall

Our own awareness is obvious whenever we open our eyes, touch an object, smell an odor, hear a sound or feel pain. The awareness of others is implied and recognized but is beyond a direct connection. We cannot feel another's pain or know, for certain, that we perceive exactly what is being perceived by another animal or individual. We connect indirectly through observation, language, and empathetic emotions. By using group contacts, we build social awareness and relationships that improve our aware connections. We nevertheless remain isolated because organic awareness is dependent upon individual neurology and physiologies. Without eyes, we cannot see. Without ears, we cannot hear and without the bundling of communicative neurons, we cannot interpret or remember the information our sensory organs provide. The connection between awareness and the physical organs that produce it is absolute. As long as the physiology is active, awareness is present. Damage the physical system and awareness is altered. Destroy the physical organs of awareness and awareness ends.

This complex dependent relationship between a genetically selected physical state and an observationally aware state is the result of mutual emergent natural selections. The ability to retain and recall relevant observations can be traced in neurological systems, and the importance of retention and recall is evident in all advanced life forms.

The evolutionary step from instant reactivity to reactions directed by stored information requires empty neural pathways to grow in conjunction with programmed pathways for reflexive reactions. When filled with sensory inputs traced in concert with reflex actions these pathways begin to offer alternate or modified reflexive actions and eventually, considered responses. When a modified recalled response is successful, the pathway is reinforced. When the response is unsuccessful, the pathway is rewritten or atrophies.

The production of neurological activity requires energy. Awareness is not free. Awareness requires a supportive physiological system capable of gathering nutrients from the environment and turning them into energy for cellular activities. Living entities with a greater capacity for neurological activity must produce a greater amount of energy and devote a larger percentage of the energy they produce to maintaining their aware state. The energy required to sustain awareness in arthropods is much less than in most chordates, but is produced in a similar fashion. The energy requirement for establishing recall pathways in our own brains becomes evident when we become exhausted by an intense learning experience. The energy required to operate the human brain is roughly equivalent to that required to operate a sixty-watt bulb or a small laptop computer. The human brain makes up only 5% of the human body by weight but even at rest, uses 20% of all the energy produced by the human body. When active it uses even more.

We have an advantage over our fellow life forms in retention and recall because of an overdeveloped brain and because of our discovery and development of a complex language. Our visual and auditory retention capacities have been supplemented and reinforced by verbal and written codes and, as a result, our retentive capacity for both detail and generalized information has been enhanced. The total amount of information in all formats retained in an adult human brain far exceeds the information stored in most libraries. Savants and others with neurological exceptions hint at an even grater potential retentive capacity.

We share memory as a general attribute with nearly every other living form including plants, but memory has its own genetic trace and is as diverse as the physical forms it occupies. We make comparative estimates of intelligence between species by measuring how quickly a task can be mastered and recalled, but our comparisons are measuring only a very small part of a much larger spectrum and we overlook many aspects of memory. Two basic types of memory are, *procedural,* the ability to recall and duplicate an action, and *declarative*, the ability to recall past aware events. Two basic methods of establishing memory are, *ingrained,* learning from repetition, and *implanted,* learning from a single experience or observation.

Two basic categories of memory are *generational,* passed from parent to offspring by genes or memes, and *individual,* an isolated memory specific to a single individual. We are grossly oversimplifying our comparisons of different species by using simple "task learning' experiments.

Awareness malleability

Problem solving is not unique to Man. Being able to manipulate retained information appears to be a natural consequence of being able to retain and recall sensory information. Individuals can learn by observing and mimicking the behavior of others, but they can also learn from their own internal repertoire of observed options. Recalled pathways of similar or related events appear to be collated naturally by neurological retentive systems. When a new input is evaluated, it is directed to paths used by similar inputs. As a new input follows collated paths and adds new information it makes new cross connections. Creativity is the recall and response activity created by these cross connections.

Nest and web building, the creative use of discarded shells for protection, the use of tools, and many other creative activities, are the result of collated yet random internal synaptic connections. Learned and repeated, these activities became permanent pathways ingrained as instincts and are supported genetically. All of this is the result of cooperative natural selection. Awareness and physical form are mutually supportive in creating survival advantages and develop in concert.

Nature's creative methods depend on a nearly unlimited number of unconditional experiments, the creativity of living forms reflect nature's methods. The ongoing creativity of both genetics and awareness is essential to life's ability to adapt to changing environments. Without malleability, life could not exist. In mankind, the manipulative capacity of retained information has reached extreme levels because of the extreme complexity of retentive pathways and because of the additional level of coded retention and recall created by a formal language.

We pass the essential concepts of science and technology that support civilization to successive generations through memes coded in specialized vocabularies. Before an essential concept can be passed on and continue to be useful, the vocabulary and relevant attached concepts have to be learned. Similarities between newly learned information and previously stored information create unintentional synaptic cross connections that act like commas in a sentence, prompting us to pause and look at the content of crossing paths during recall. These alternative pathways are at the root of our ability to manipulate recalled material. Manipulated recall is also responsible for our creativity. These unintentional neurological intersections are at the root of scientific research, mathematical experimentation, and innovative responses. Manipulative awareness adds a new level to natural selection, a level beyond randomly altered repetitive awareness.

Language

From squids to insects, to birds, to man, language adds an adaptive mechanism for the establishment of communities of awareness. It also facilitates the transfer of bits of experience from one aware form to another, uses many different media, and has many forms. Language, as a naturally selected attribute, has played a major role in the evolution of awareness. (See chapter 11)

Pools of awareness

As an option to living as a mass of immortal sister cells dividing endlessly, naturally selected biological forms incorporated, *individuality*, (being slightly different from everyone else), by developing sex, (gene swapping), and death, (discarding cells after gene transfer). The unintended choice by nature to incorporate sex and death occurred as a way to expedite genetic variance and produce a larger pool of physical options to test against changing environments

Awareness has followed genetics in this preference for individuality out of necessity but as physical forms continued to adapt, individual neurological complexity began to reach a limit. To overcome this limit nature began experimenting with social awareness. The value of social cognition, as a supplement to the limitations of individual awareness, has fostered the natural selection of both complex languages and complex social groupings. Insects coordinate great colonies with subtle chemical languages, marine animals communicate with body language and florescence, and mammals, including man, form social cognitive groupings using a variety of vibrational languages.

Social cognition is made possible by language and by swarm intelligence. Group intelligence has great survival value. The concept of a meme (pronounced *meem*) as the unit of transfer for acquired information, and as a partner to the gene, can be helpful in understanding these social interactions and in the analysis of most methods of information transfer. (See Chapter 6)

Individual awareness is limited by genetic, sensory and neurological limitations. The relationship between physical organs responsible for awareness and awareness itself is absolute. The capacities of one create the limits of the other. Shared memes create meme pools that improve individual limitations, but are indirect and, except for mankind, do not have permanence beyond individual retentions. Meme pools are also dependent upon individuals for the addition of content and are of value only when referenced by individuals. Social meme pools remain empty until filled by the experiences of individuals and silent until an individual dips into them for information.

Meme pools are carried as shared information only within life forms with an active awareness, and are limited to individuals comprising an active social group. In man the invention and use of symbolic language has allowed meme pools to transcend time without the necessity of being carried by an active individual awareness. Symbolic written meme pools are also silent until referenced but can grow beyond the retentive capacities of individuals and are no longer dependent upon the continuing transfer of information, or by mortality limits. Symbolic pools are also less susceptible to the morphing of original meanings.

The static nature of mechanically recorded memes protects them from the imperfect and often faulty chains of transfer required for active awareness pools, but, by being static, the meme pools of man are also isolated from the normal pace of natural selection and create groups dependent upon social memes that are no longer relevant in current social or natural environments. When a meme pool fails to evolve because it has become isolated from, evolving determinants, the demands of natural selection continue and can declare the pool obsolete. Social groups dependent on obsolete meme pools may struggle to ignore natural selection through denial and intensive restricted educational transfers, but the process of natural selection will eventually eliminate or adjust all out of place genes and memes. Man adds impetus to the process of natural selection, but cannot deny it or defy it.

With the invention and use of communicative tools, an indirect augmentation of social awareness has been created. The telegraph, the radio, the telephone and now the internet have accelerated the growth of man's social awareness by making social information immediately available to more individuals at greater distances, but true awareness still resides solely within the individual. The capacity of our most powerful computers approaches what could be considered awareness, but in spite of the tremendous amount of information stored and their extreme computational capabilities, computers remain comparable only to the passive aware states of more primitive biological forms

Active awareness, cognitive awareness and self-awareness can be demonstrated in small isolated digital experiments but only at levels that have little in common with living awareness and remain well outside self- advancement through natural selection. Replacing language as the primary transfer agent of memetic information with a mechanical direct connection between individual aware states is still far beyond our capabilities and probably beyond neurological capacities as well. The genetic natural selection of increased neurological capacity in individuals may or may not be taking place in current human mating selections, and gene manipulation to create individuals with advanced capacities has risks and moral implications that will preclude its use in the near future.

Our ability to use our innate neurological capabilities more effectively, to cooperate in intellectual endeavors in ever widening arenas, to use improved information technology and mechanical devices to augment our cognitive abilities, and to use our ability to transfer complex memetic information, continues to widen our social meme pool. Until now, we have been able to accommodate this widening social pool without exceeding our individual neurological capacities, but a limit will, eventually be reached.

The specialization needed for individual meme sharing and retention will continue to increase as the complexity of our social meme pools increase and will eventually surpass our ability to coordinate the diversity.

Expanding meme pools are filling rapidly, and will soon require additional individual specialization for their content to remain within individual capabilities. All of physiology and medicine once fit into an individual doctor's synaptic capacity. Now the medical meme pool requires many specialists acting as a team. Specialization increases meme transfer requirements and creates further fractionalization. The number of specialties required in some fields may soon reach coordinative limits. Natural selection however has stepped over many hurdles before. Should humanity fail, nature will surely find other ways to awaken the universe.

CHAPTER SIX

Chapter Contents

CHAPTER SIX

GENES, GEMES AND MEMES

Neutral and natural selection

NEUTRAL genetic selection occurs in genetic transfer when a misplaced allele or codon signals a variation from the original message that has no immediate effect on the resulting life form's ability to survive or procreate. Coding for neutral characteristics can be carried forward through generations and remain inert until a situation arises that gives it significance. A similar type of neutral acquisition occurs when a MEME, (a bit of information passed by language or observation), is assimilated but serves no immediate purpose and prompts no significant responses until a situation arises that gives it significance.

NATURAL genetic selection occurs when a misplaced allele or codon signals a variation from the pure message that has an immediate effect on survival or reproduction. "Immediate, in evolutionary time scales, can mean dozens or hundreds of generations during which the statistical advantage or disadvantage of the new trait is tested against competitors or the environment". A similar natural selection occurs when a MEME (the unit of transfer of awareness) prompts an action or a change in behavior that has a significant or lasting effect on survival or reproduction.

The Gene

(The unit of transfer for form and function)

Genes pass on their instructions for recreating a single cell or a multi-cellular structure as a coded set of instructions. The human genome contains over two million alleles. Among these many sets, special alleles called codons signal RNA to start and stop the transcription of proteins for specific physical attributes. Genes are the alleles located between these punctuation marks. The human genome contains 23 chromosomes, (sentence strands), about thirty thousand genes and two million alleles, (individual amino acid letters). A large percentage of the genomic material appears to be surplus and nonessential. These surplus strands may be leftovers from previous genetic experiments, random alleles, or serve a purpose yet to be discovered.

The Geme

(The unit of transfer for reflex and instinctive behavior)

Between the active genes on the genome ladder are millions of allele repeats and random arrangements like stuttering in the language of life. We have not found a purpose for these long ramblings but we also have not found the codes that hardwire preferred synaptic paths for reflex and instinct. Darwin spent considerable time documenting instinctive and reflexive behavior but lacked any scientific evidence to explain it. We have progressed little in this area and although we are now convinced that there is an as-yet undiscovered physical mechanism for the transfer of instinctive and reflexive information, we have not been able to describe it in detail.

The vast neurological networks common to all multi cellular life attest to the importance of a system capable of interfacing and reacting in a timely manner to environmental inputs. We can trace long mono-polar neurons from receptors in various parts of the body to more primitive parts of the brain and identify their activity in reflexive reactions. Shorter connections within the brain, between neural pathways and motor neurons that cause directed actions have also been identified but are less well mapped and less well understood. The human brain has approximately 100 billion nerve cells and the human body contains an equally impressive number. The production, positioning and (initial connections) between neurons appears to be generally directed by genetic coding, but not specifically directed. Non-coded natural patterns, such as branching, may also play a part in non-specific neural arrangements.

Genes code for the intricate neurological interfaces between sensory neurons and reflexive motor neurons. Genes also code for the more complex wiring of generalized and cumulative sensory inputs, emotional response centers, and preprogrammed complex behavior centers. Specific genes responsible for the development of various brain cell centers and their connections are being identified in organoid brain tissue grown in laboratory conditions but more study is needed. To separate the alleles in a genome responsible for pre wired behavior from those responsible for other physical characteristics is difficult, but for our discussion, we can separate them by calling genes that program for behavior, Gemes, and may find them more prevalent in life forms that undergo a metamorphosis.

The Meme

(The unit of transfer for acquired elements of awareness)

The unit of transfer for repeated behavior is the meme, originally called the mime, (a concept with limited applications). The author has chosen the spelling and definition that identifies the meme and memetics more closely with the gene and genetics in order to broaden the concept of awareness beyond the more limited concept of consciousness.

We have known that behavior can be transferred, (learned), for thousands of years but did not give much thought to the process or its origins until we discovered the gene. If physical characteristics are coded for transfer in a chemical language, is behavior similarly encoded? If genetic transfers facilitate evolution from simple forms to the more complex, do memetic transfers facilitate an evolution from simple behavior the more complex.

We can observe the gene, and the results of its RNA directives. We can see the gene and measure and manipulate it. Genetic functions fit well into our scientifically based perspectives. The role the gene plays in evolution is clear. Awareness and its evolution, on the other hand, are not as easy to examine. We can observe and measure awareness, categorize its patterns, and even manipulate it, but we can't touch it. We are certain awareness exists because we, as individuals, cease to exist if we loose it and because we observe it in other living things. We cannot deny the existence of awareness but have trouble describing it, and as a result, have often given it mystical and religious properties. If we view awareness analytically however, awareness becomes a naturally emergent characteristic of complex organic arrangements. From this perspective awareness, including the awareness of man is still mysterious, but only as one of nature's many combinational miracles. Awareness, as a natural by-product of living matter becomes a full partner with the gene. As genes test physical attributes against the environment, memes test learned responses against the same environment and the combined testing results determine the future shape and responses of creatures with advanced awareness.

The separation and duplication of the genetic double helix may be evidence for awareness at life's most primitive level. The genetic double helix is a physical entity but its separation and duplication is a prompted behavior, a simple chemically prompted form of awareness.

Memes, units of learned information and behavior, are passed to future generations, not by sex, as is the gene, but by example or instruction.

Memes are created by observation as one individual notes a successful or unsuccessful behavior, or by a personal experience with positive or negative results. Memes can be transferred one individual to another or one individual to a group and like the gene can have their instructional value modified during transfer and over time. Meiosis, sex, accelerated the gene's ability to create and test new living forms just as language accelerated the meme's ability to create and test new ideas and responses.

Meme transfer cannot occur until genetic selection has first produced a combination of sensory and neurological retentive abilities capable of accommodating memetic transfers. In primitive life forms, meme transfer is infrequent or null and plays a secondary role to genetic selections. When a gene begins to code for a neurological characteristic that allows for even a simple meme response with a survival advantage, the gene coding for the memetic advantage is placed on the genetically preferred list. With each gene selected as advantageous because it codes for memetic advantage, the partnership between genes and memes is solidified and, functioning in unison, both begin to code for advances in awareness.

The meme like the gene cannot exist separately from its host. Memes must be passed forward or they disappear. Memes like genes are modified by circumstance and are adjusted or replaced as circumstances change or when their survival advantage is lost.

In living forms with advanced sensory organs and well-developed neurological centers for coordinating information and activity, meme transfer is rapid, impacts survivability immediately, and can override genetic selections. Use memory to respond appropriately to observe and recall a response by another individual or remember a personal event previously encountered, and the meme overrides genetic selection.

When a meme overrides the gene it usually supplements the gene's survival value, but it may also allow a gene, out of sync, with the environment to carry forward, dependent on the meme to protect it from extinction. The partnership between genes and memes can be destructive, supportive, or protective. Supportive and protective relationships result in continuing genetic coding for support of the meme and further accelerate coding for advances in awareness.

The meme as the genes full partner

Genes are replicated to insure the continuance of valuable physical characteristics.
Memes are duplicated to insure the continuance of valuable animations and responses.

Physical form is of little value without appropriate animations. All living things need to respond appropriately to changing environments if their genes are to survive but genetic responses are slow and not all behavior responses can be genetically encoded.

Nearly all of the thousands of species ever evolved are extinct because their environment changed too rapidly for genetic drift to accommodate. Most of these extinctions were mass extinctions caused by global catastrophes but between these events, many more species were eliminated because life's method of offering genetic options as random test for survivability were too few and too slow. With the introduction of genetic preferences for advanced awareness a survival supplement, the meme, appeared and working in partnership with the gene created new survival options.

The gene creates the opportunity for memetic transfer and meme transfers help to insure the survival of the gene.

The gene has evolved from a short strand of replicating amino acids into hundreds of cooperative strands with millions of alleles.

The meme has evolved from primitive learned responses to complex signaling and languages with hundreds of symbolic forms and has coalesced into memetic groupings to form millions of concepts.

Meme transfer takes place from individual to individual, from an individual to a group, (and with the advent of territorial marking and writing), has the ability to transcend time between transmission and reception.

Gene transfer also takes place from individual to individual and (in the case of external fertilization in marine environments), from an individual to a group and in the case of dormant or inactive genes, can be carried forward to transcend time.

An animal giving a warning call passes a meme alert immediately to all who can hear the call. A historian reading Homer in the twenty first century is receiving memes delivered centuries earlier.

Memetic evolution appears most influential in the process of natural selection leading to man but the genetic predisposition for mimicry is evident in most of our primate relatives. The gene meme relationship responsible for mimicry has passed millions of natural selection filters and originated far back in the history of life. Mimicry (the ability to reprogram one's synaptic response repertoire through meme transfer) began early, long before the first proto-primate. As improved sensory and neurological capacities facilitated the ability to find food, avoid danger and find suitable mates, the ability to mimic successful behavior was also naturally selected, and as genes continued to be selected which favored better observational and synaptic capabilities, more complex arrangements of memetic information became capable of transfer. The meme, like the gene is only a symbol, a code that needs to be translated into action or a physical form. Genetic codes are sequenced arrangements of amino acids ready to be translated into proteins that become living reality. Memetic codes are retained observations read by organs of awareness, recorded as meaningful groups of stored synaptic arrangements (Themes), ready to be translated into useful responses. Inherited physical forms and acquired themes are the reality resulting from these two types of coding. These two aspects of life determine how we look and how we act.

It is interesting to note that both living forms and the behavior of living forms originate from, and are sustained by symbols, chemical and electrical.

Memetic modifications and transfer

Changes in the environment adjust genetic material and memetic responses by testing the success of creatures produced and directed by their directives. Code for the production of a creature better adapted to a new situation, and the code persists. Code for a creature not as well suited and both the creature and the code are eliminated. There is no advanced planning and no direct reactive adjustments. Genetic material drifts slowly and creates duplication errors occasionally, but it does not think or make intentional adjustments. The success of a genetic adjustment is always serendipitous at the individual level. At the group level however, statistical probability provides a better assurance of success. Cast twelve dice and the probability of four sixes is low. Cast a hundred dice and the probability of four sixes is high. Add complexity to a creature and the statistical probability for genetic success improves. Add the ability to make immediate reactive adjustments to changing situations and it improves even further. Awareness adds the ability to make immediate reactive adjustments. It is present in all living things and as it became more useful, it has adjusted the genetic natural selection process to further its presence and enhance its effect.

Environmental changes select new genetic codes only if new codes become available. Environmental changes do not create new genetic codes directly. New genetic codes depend on genetic drift or gene faults.

Changes in the environment can however create new memes. The sudden onset of cold weather cannot immediately cause a creature to grow more fur unless some of the creatures already contain coding for a warmer coat. The onset of cold weather can however cause creatures suffering from the cold to use their advanced awareness to seek relief by modifying their behavior, and if one of them discovers a way to keep warm, it creates a potential survival meme available to the group.

Unlike the gene, the meme is not restricted to transfers within a species. Memes can pass on a survival advantages without deference to breading compatibility. A species observing a successful memetic response in a different species can add the meme to their own awareness pool, incorporate it as a survival response, and pass it on to others by example or instruction. The meme exists as a set of stored neural synaptic information just as fur exists as the product of genetic alleles. The gene imparts structured arrangements through chemical coding. The geme imparts responses by creating favored genetically selected neural pathways. The meme enables complex reactions through learned synaptic alternatives.

Memetic evolutionary theory

Memes, *(The units of transfer for learned behavior and information)*, are created in response to genetic expressions and their host's interactions with the environment. They are modified by circumstance, observation, instruction and, (like the gene); by occasional faults in transmission from individual to individual. Recorded, stored and accessed as. synaptic connections, memes, (like genes), are tested against the reality surrounding their hosts. Successful memes are used repeatedly and passed on to offspring and following generations. As circumstances change, new behavior patterns of relevant information, (memes), are added and old synaptic connections are set aside. The process is proactive, generations faster than genetic modifications, and has its own evolutionary history.

We can trace genetic evolution back through millions of species to its early origins by identifying common individual genes incorporated in successive life forms. We share common genes with bacteria, molds and plants and although the lineal trace is difficult, the history of genetic development is available in a written form.

Tracing memetic evolution is more difficult and may not be possible with any assurance of accuracy but reasonable generalizations can be made.

Like common trace genes, aspects of acquired behavior also pass from individual to individual and from species to species and can be isolated and traced. Memetic evolutionary history is much shorter that genetic evolutionary history and, although evident in more primitive kingdoms and phyla, became evident and influential only with the advent of animals.

Identifying behavioral and neurological patterns common to living forms sharing an interactive past, exposes a chain similar to a genetic trace. Plotted, these common memetic patterns expose a memetic evolution working in concert with genetic evolution.

It is important to separate memetic transfer from the gene's mechanical transfer methods to appreciate memes as evolved, stored, and shared, units of information, The gene and the meme hold hands in the evolutionary process and both influence natural selection but in different ways. In primitive life forms, physical attributes are the result of genetic adjustments acquired through many generations. Memetic adaptations in primitive life forms are rare, are acquired through fewer generations, and are coupled with genetic changes in ways that makes their evolution difficult to identify. In more advanced living forms, memes play a larger role in evolution and genetic and memetic methods take on separate identities.

A bird of pray that inherits a dimple on its cornea allowing it to focus on a distant target using telescopic sight, has a genetically acquired advantage that, over time, is carried forward to future generations. A similar bird of pray without the telescopic advantage, but with advanced observational skills, learns through repeated observations that the advantaged bird repeatedly hovers and then makes a definitive circular attack to catch its prey. The watching bird learns that the genetically advantaged bird's circular flight path always points at a spot in the grass where unseen prey should exist. After a few trials, the bird with normal sight also learns that it can dive on the invisible target ahead of the telescopic hunter by taking a more direct path.

The naturally selected advantage of telescopic sight is a genetically evolved trait. The new synaptic arrangements created in the brain of the observant bird are memes. The genetic advantage passes to following generations by sex. The memetic advantage passes forward through observation and by instruction. Both advantages are tested by natural selection and, in this case, the memetic advantage takes precedence over the genetic advantage and the bird without telescopic sight has dinner compliments of a genetically advantaged bird that has now become the memetically advantaged bird's spotter. For the time being, the memetic advantage gets the prey. Over time however, the telescopically advantaged bird may learn to adjust its hunting behavior to outwit its shortsighted competitor and add another meme to the mix.

In the scenario above, the influence of memetics has temporarily trumped genetics and the evolution of both species of birds has been altered. The memetic hunting techniques of birds of pray can be inferred by observations but can never be as definitive as a genetic trace of physical hunting attributes. The gene is a written record. Memes, on the other hand, exist only as a temporary synaptic arrangement and interpreting these arrangements is possible only by observing current individual and group behavior patterns.

Genes transfer as physical units. Memes transfer as instructions or suggestions. The smart bird sets an example for others to follow, but following it is always an option. Following the results of a gene transfer are not optional.

In more complex living forms, with more advanced neurological capacities, memetic transfers become more prevalent and synaptic patterns are copied more accurately, (especially as languages in land and marine animals evolved). To explore memetic transfer and its evolution several scenarios can be used but examinations can begin by observing a broad range of phyla and species looking for similar memetics in their history and any common genetic adaptations associated with their memes..

Enhanced natural selection by gene/meme coupling

In advanced life forms, memes transfer behavior and genes transfer structure and interact in ways that can be;

- independent
- counter productive
- coincidentally cooperative
- mutually perpetuating

When genes and memes cooperate, the effect is powerful and can offset many of nature's basic guiding principles. When they become mutually self perpetuating they can override natural selection.

Genes are passed from individual to individual only through the passing of genetic material from parent to offspring in a sequential chain. Adaptations are slow and spontaneous genetic changes are rare. Faithful genetic reproduction fits effectively with slow changes in gross environmental conditions but cannot keep pace with rapid changes.

Memes are passed from individual to individual and from individuals to groups by observation and instruction without regard to familial genetic chains. Adaptations are rapid and can accommodate rapid environmental changes and spontaneous memetic modifications are common. Misinterpretations of observation and instructions occur frequently.

An example of *independent* effects caused by the activity of genes and memes is genes drifting to code for thicker fur, while memes in the same species pass learned instruction for improved hunting techniques.

In this example, the gene and the meme influence the survivability of the species independently.

An example of *counter-productive* effects caused by the activity of genes and memes is genes drifting to code for nearsightedness while memes, (units of learned behavior), direct the flock toward a new pray that requires detection at a distance.

In this example, a developing physical attribute is acting in opposition to a newly learned preference for new source of food.

An example of *coincidental co-operative* effects caused by the activity of genes and memes is genetic drift, to code for feathery gills increasing oxygen uptake while learned behavior is directing the creature to forage out of the water for food washed on the beach by the tide.

In this example, the genetic instructions and learned instructions compliment each other, and prompt the first animal to venture onto dry land.

An example of *mutually perpetuating* genetic and memetic influences is man. The species of man is the result of a fortuitous juxtaposition of genetic anomalies with learned behavior patterns that promoted the complimentary union through genetic selection and improved survivability.

Homo sapiens are the result of favorable genetic drift occurring simultaneously with the development of valuable behavior patterns.

Other species with advanced characteristics resulting from genetic/memetic perpetuating pairs exist, but none with the number of simultaneous pairs that prompted the evolution of man. The evolution of man is the result of four perpetuating pairs of genetic and behavioral tendencies.

1. Genetic drift toward upright walking paired with behavior memes encouraging less time in the trees and more on the ground,

2. Genetic drift toward more grasping capable hands paired with observation and instruction memes encouraging the use of simple tools,

3. Genetic drift toward increased neurological capacity paired with example and instructional memes encouraging curiosity and increased alertness,

4. Genetic drift lowering the vocal chords in the human throat paired with example and instructional memes encouraging articulate vocalizations.

The combination of slow genetic evolution with rapid memetic adaptive capabilities is a very effective survival pairing, especially when memetic patterns persist long enough for genetic modifications to enhance wide pools of memetic information. The combination accelerates adaptive changes and, in the case of man, can override natural selection.

The most advanced neurological system is now paired with the widest pool of memetic information in an animal known as man. In this creature, the advanced combination of neurological capacity and accumulated information is overriding the process of natural selection. Instead of the environment modifying man to fit the environment, man is modifying the environment to fit his percieved needs. Memetics has usurped genetics by combining an advanced neurological capacity with a prodigious meme pool.

Man has temporarily, suspended the rules of natural selection as they apply to his species, and influences the genetic future of all other life forms on the planet. The memetic control wielded by man also allows him to by-pass natural selection by altering nature's language directly, shaping attributes and mutations through manipulations impossible in nature.

Unfortunately, man is incapable of forecasting the outcomes of his manipulations and dominance. He claims supremacy, but natural selection will always be the final arbitrator. In man, natural selection has been suppressed, but it is only 'on-hold' while a new approach is tried.

Themes, concepts and perspectives

Early memes survive in man as synaptic capacities that mimic the patterns and processes that preceded them. We think in patterns similar to the way nature is organized because, we are the product of nature. The storage and computational capacity of the human brain however, is far greater than that needed for basic survival responses, and has been supplemented by its own inventions, especially language. The power of the human brain to observe in detail, decipher complex arrangements, and invent ways to alter the environment to suit its own needs, appears to be an evolutionary over step. No longer is life just passively adaptive. The power of the surplus synaptic capacity in man's evolved brain now actively controls outcomes once reserved for mechanical processes. Simple memes have been gathered by man's synaptic excess into relevant concepts and complex themes, like governments, religions, and economic structures. As man analyses and reorganizes his stored memetic information, he imagines situations that have never occurred, sounds that have never been heard, and matter organized in ways that natural processes could never accomplish. He organizes nature into agriculture, organizes himself into social groups, creates music and art, and turns natural elements and processes into architecture and machines.

Further enhancing his evolutionary dominance, he selects, gathers, and compares his synaptic excesses to create concepts and groups of concepts to form themes and uses his complex and flexible language to symbolize and share his larger ideas to create social networks that become established patterns and create perspectives that direct and control individuals, and groups.

Social anthropologists have made many studies of languages, social organizations and the use of tools as comparative markers, but the true nature of any human social group is carried in the group's common concepts and perspectives. Accessing these is difficult because they exist only temporally as individual synaptic arrangements, and are usually a mix of information and emotion. Individual and commonly held concepts are reflected in behavior, in persistent myths, and in the vocabulary of a group's language. The power of the meme in man has usurped the gene in the process of natural selection but cannot avoid the scrutiny of the natural selection process.

Man's future is still completely dependent upon a successful fit with future environments, natural or manmade. What is unique to man, after four and one half billion years of genetic evolution, is his current nearly total dependence upon memetic evolution. Man's memetic capacity has evolved into complex themes, concepts and perspectives that override most of his dependence upon genetic selections.

Humankind's fit with the unforgiving forces of nature, now depends upon his synaptic arrangements, his emotions, successful social concepts, and realistic perspectives. We not only have to conceive the correct path to the future, we have to agree upon it and then follow it. Diversity of thought is both our enemy and our ally. Among the evolving diversity of concepts and perspectives successful answers may exist but getting past our divisive tendencies may be impossible. If the history of humankind is a reliable indicator, we will continue to kill each other over competing perspectives, render many concepts extinct, establish many changing dominances, create or encounter environmental changes that redirect our perspectives, and may, if we continue to deny reality, negate man's synaptic excess as a viable survival alternative and revert to a genetically dependent animal. We are now responsible for our own natural selection, or de-selection.

The key to our future lies in the order and precedent of our conceptual groupings. We have the power to insure our future by creating common viable perspectives, but we are still waging conceptual wars between the obvious around us and imagined unseen deities. The alternative to agreement is to continue our concept wars between 'belief' and 'being' while human disasters continue to provide believers with proof of a vengeful god and non-believers with proof of the impersonal process of natural selection. As long as both sides are using the same evidence as proof for their perspectives, any intellectual resolution remains impossible and we will continue to compete and wait for God or Nature to decide our future.

By default, natural selection has put us in control of our own evolution. We have learned how to accelerate the process but we have not learned how to guide it. We are riding in an unguided vehicle arguing about who should steer.

CHAPTER SEVEN

Chapter Contents

CHAPTER SEVEN

MEMETICS

It may seem presumptuous or even superfluous to create new categories and titles for well-studied academic disciplines, but by examining cognizance and intelligence from a new perspective we can leave behind words, concepts and symbols that have become overused and limit our view. Seeing awareness as a natural and essential reality present in all living things allows us to look beyond our own reflection and step away from established preconceptions. Philosophy is the art of creating useful insights and perceptions and has practical value when it prompts alternate intellectual excursions into established areas of inquiry by expanding limiting concepts and vocabularies.

Genetic/Memetic interdependence

When life was new, genetic drift was the primary producer of new characteristics. As advances in form and reactivity were tested, complex sensory and reactive capacities were naturally selected as compliments, and eventually allowed life to adapt to living on land and to moving through the air. Any pilot, accustomed to making landings on wide paved runways, cannot help but appreciate the advanced state of these closely joined genetic/gemetic/memetic interactions, as birds adeptly land in a tree. The ability to spot and evaluate one small tree limb among hundreds as a suitable landing spot and then to make a high speed approach avoiding other limbs by flexing wings and adjusting control feathers in rapid succession to arrive from a proper direction and at the exact speed needed to accomplish a successful landing. The approach of a bird to a landing is a very complex activity and requires a very advanced state of awareness. This feat and thousands more just as complex are the rule rather than the exception throughout the biosphere. In the above example, sensory neurons continually transmit information regarding pressure on the wings, overall orientation, and velocity information from airflow across speed sensing feathers to neurological feedback loops that flex small finger feather muscles to control roll and pitch and larger muscles to position wings and feet. All of this activity occurs simultaneously in a continuous flow of visual inputs and muscle positions that are instantly analyzed by a central neurological center where active choices are added that redirect the complex loops in milliseconds and send adjusted control signals back along motor neurons. The process is a continuous loop of sensory input, analysis, feedback and control. Sensors, muscles and neurons are all genetically produced, but much of the activity they support is learned. Learning takes place when semi permanent synaptic connections are established from the repetitive use of input/ control loops and, over time, are imposed as hard-wired connections. This overlay process is a mix of mimetically acquired information and genetic hard wiring and is evidenced in all creatures with an active level of awareness.

Most complex behavior is a mix of genetic preparation and memetic acquisition. Complex tasks can be learned only if the capacity to acquire and store the necessary information is present, and can be activated only if the physical form has been appropriately prepared. Physical attributes and the ability to acquire skills are naturally selected as supportive pairs. They develop simultaneously prompted by their complimentary advantage. A physical form with great strength or great speed has no advantage if it lacks the neurological capacity to direct its physical advantages toward relevant outcomes. The natural selection of these combinations begins with gene selection but as neurological capacities prove mutually beneficial, memetic capacities are added. The process is mechanical and random, a simple and continuous testing of muscle, bone and neural combinations. Successful combinations continue. Those that are not successful are eliminated.

Awareness advances as a natural by-product of the growth of neurological systems when neurons are positioned by natural selection to accommodate favored synaptic alignments promoting useful instinctive (gemetic) behavior. Gemetic behavior is an interim functional state between reactive behavior, (reflex) and active awareness, (choice making). These gemetic arrangements are the prototypes of neurological programmable circuits that allow an environmental interface one-step above passive awareness. Gemetic behavior occurs when immediate environmental assessments begin to cause slight modifications to strict reactive behavior by recording these modifications as preferred synaptic arrangements. These new, preferred, synaptic arrangements then become a part of future genetic selections. We recognize these semi programmable behavior patterns as instinct and they display themselves as both simple and complex behaviors ranging from the coded flash of a firefly to the long migrations of fish and birds.

Complex instinctive behaviors are more malleable, allow input from active awareness, and begin to introduce choice. Being able to choose a response based on an immediate assessment of one's surroundings is a valuable survival attribute and creates an equally valuable feedback pattern for natural selection. Neurological arrangements associated with active awareness expanded and improve as instinctive behaviors are naturally selected to include choices between multiple response options.

The value of active awareness to impact behavior immediately is what gives it precedent in natural selection. Migrating birds that cannot adjust their flight path or delay their migration to avoid a winter storm put themselves at risk. Those that can assess their immediate situation in more detail and alter their behavior accordingly, are more likely to produce offspring with similar aware skills. Advanced levels of awareness fostered by advanced synaptic capacities for both information acquisition and relevant recall; add survival values that are more likely to pass through natural selection's filters.

Memetic categories

Memories are small packages of acquired sensory information that are stored as lasting neurological synapses. The duration of these recallable synaptic links can be short or last for the lifetime of an individual. They can also become self-sustaining and exist beyond individual lifetimes by becoming a meme, an acquired behavior or observation that is transferred sequentially between individuals. Memes can last for hundreds of generations but, like the gene, the meme disappears if not transferred. If a meme is not passed on, it becomes extinct. Genes pass on physical characteristics, and synaptic predispositions, while memes pass on information and alternate patterns of behavior. The gene passes to the next generation, as a coded order of amino acids. The meme passes to the next generation by sensory acquisition, and is recorded as a synaptic trace. Describing the act of learning as the transfer of memes may seem illogical until one realizes that the gene too, has no visible hint of the form it is about to produce. The reality, form and behavior, that the gene and the meme carry forward from generation to generation, is not evident in the unit of transfer. The gene we see with a powerful microscope has no resemblance to the life form it describes. Likewise, we can witness a learning event when a meme is passed forward, but not the synaptic trace that carries it forward. Shape and function are both passed forward as distinct units in a coded form, one coded in amino acid arrangements, the other in electrically induced synaptic arrangements. Thinking of the transfer of memes as the passing of units of evolved behavior allows us to understand how awareness can be naturally selected as a primary survival element, and how meme pools can create group awareness. The mandate for the gene to code for physical form is paralleled by the mandate for memes to code for living experiences. Human civilization has been created and is sustained by meme transfer and the many cultures of humankind can best be understood by isolating and analyzing their meme pools. Studying and understanding the common bits of information that sustain and shape a culture helps us understand both the behavior of individuals within the meme pool and the behavior patterns of the group.

In its most basic form the meme is the synaptic retention of elemental sensory inputs that make up an observed event. The event can be *static*, such as the making of a synaptic recording of the sight, sound, smell, and feel of an observed place or thing, or the event can be *active*, such as a synaptic recording of the interaction of various elements. Memes can be simple, as in the learned recognition of an object from a simple single sensory imprint, or complex, as in a learned response to a situation from a complex synaptic imprint of multiple sensory inputs. Meme prompted behavior depends upon timely and directed access to stored memes whenever a later sensory input triggers associations through neurological meme searches and meme recall.

Memes are sensory acquired bits of information stored as synaptic arrangements that prompt behavior or additional recalls. *Immediate need* prompted behavior is initiated by internal or external sensors that trigger hard wired, genetic, synaptic pathways that result in glandular secretions of adrenalin, hormones, endorphins, etc. Chemically responsive synaptic pathways are primitive response mechanisms that produce urges such as hunger, fear, and sexual activity. Primitive pathways have been retained in more advanced life forms as essential passive aware mechanisms by developing connections to advanced active synaptic pathways that allow sensory inputs to modify or redirect their responses. These modified, combinational connections between primitive neurological centers and more evolved centers are a mix of genetically created pathways and memetically acquired pathways.

Memes, like genes can be neutral or active. Neutral memes are memory bites that are seldom if ever used and do not impact survival or reproduction. Active memes are referenced often and their synaptic imprints are reinforced with use. Memes are bits of information acquired through the senses by an individual life form and can exist outside a synaptic arrangement, only as static inert symbols. Both neutral and active memes can be transferred by mimicry or instruction to create a similar synapse in another life form, and if naturally selected for repeated or mass transfers, can become a social meme, (something all living entities in a group know in common), or a survival meme, (something that must be known in order to survive).

An example of a neutral meme would be that tree bark feels rough. An example of an active meme would be, in the case of koalas, that eucalyptus leaves are edible and essential to sustain life. The feel of tree bark is an individually acquired meme and requires the individual to rub against a tree. One must experience rough bark on their own to acquire the synaptic arrangement that constitutes the meme. The taste and satisfaction resulting from eating eucalyptus leaves, on the other hand, is first observed, then reinforced and passed from Koala parent to its offspring as an active social meme.

The information that coral snakes are deadly and should not be touched is a survival meme. Failure to observe this essential meme can be fatal. To be of value, survival memes need to be available for immediate recall and any neurological pathways offering distinguishing comparatives need to be active. Knowing what a coral snake looks like and knowing that it should not be touched is a valuable meme, but is useless if an encounter with the deadly snake fails to distinguish the snake from other snakes and connect with the a response meme to stay away. The visual image of the coral snake must be associated, through a comparative synapse, to a stored visual synapse, and the "do not touch" survival meme activated. I call these connections, *memetic recognition responses.*

Recognition memes can be as simple as a single visual object imprint and can be neutral or active. Seeing the odd shaped rock at the corner of the path leading to the water hole every time you go for a drink creates a neutral object recognition meme. If the rock is moved, you still recognize the object but you also recognize the fact that it is in a new location and the setting recognition meme is altered. If the new setting also involves the presence of an attractive potential mate, a situation meme is created and if the potential mate wiggles its tail, a response meme is created and the rock takes on less significance.

Memes are small synaptic packages of sensory inputs. These packages can be a visual imprint of a simple object in a simple setting or a mix of sensory imprints composing an entire situation. Memes can be neutral or active. They can also be individual or shared and become part of a social meme pool. Most memes are recognition memes, stored synapses of sensory inputs in an active neurological net ready for comparative analysis and recall. Any mobile life form with an active awareness creates dozens of memetic synapses every few minutes. In pools of active awareness memetic transfer between individuals is ongoing, and as sustained memetic chains are formed, they begin to affect genetic outcomes for the group and its members. Memes are illusive and transitory and transfer only as temporary synaptic neural pathways, they are however, as real as the alleles on a genetic spiral and have an equal impact on evolution when they are naturally selected or de-selected. Language also reflects memetic arrangements and functions. (See chapter 11)

Genetic and memetic modifications by meme pools

Natural selection uses all aspects of the environment as a filter to insure life's continuance. Physical environmental conditions and processes predominated as the primary selection filters when life first began, but living forms now fill nearly every environmental niche available and are, themselves, as much a part of the natural selection filtration system as ocean salinity, sunlight, volcanoes, draughts, and earthquakes. The complexity of the living environment, like physical environments, is extreme and oversimplified concepts like, 'the survival of the fittest' only diverts our attention from the complex continuum around us. Complex biospheres, food chains, symbiotic arrangements, competition for resources and more, all affect what future generations will look like and how they will behave.

The living environment continually modifies the physical environment and the physical environment continually modifies the living environment in a tangle of events beyond measure and comprehension. Evolved awareness is a part of this complex mix and influences both genetic and memetic selections. Beyond genetic and memetic relationships at the neurological and synaptic level, hard wired neurology and software synaptic arrangements impact and interact at higher levels. Group dynamics acting in concert with active and cognitive awareness create new memes, concepts, themes, and perspectives that have their own impact on natural selection.

Similar life forms have been assembling into social groups to add survival advantage since the development of meiosis produced diploid cells. When single cells acquired the ability to swap genetic material their adaptive ability accelerated exponentially, and the survival value of grouping was naturally selected immediately. Fossil records of these earliest of living societies exist as mineralized hummocks left by the layered groups of single celled organisms that first exploited this survival advantage. In this earliest evolutionary experiment, with diverse cells living in proximity, active awareness had no impact on genetic outcomes. Passive awareness was the norm. After several billion years of naturally selected complexity in both genetic coding and increased capabilities in passive awareness, multi cellular life arose and the influence of both genetic selections and memetic selections increased significantly.

Nearly all land and sea animals, both arthropods and chordates, arrange themselves into like kind social groupings. Schools of fish, whale pods, insect swarms, flocks of birds, and human societies are all examples. Language, of some sort, is a common trait within all of these groupings. The more developed the language the more adhesive the group and when these languages morph into slightly different calls or visual signals social groupings splinter in much the same way that genetic drift creates new physical characteristics and species. These new language specific social groups isolate themselves from their predecessor groups and in doing so also isolate their gene pool. The impact of language drift on the drift of social groups directly affects genetic outcomes and natural selection becomes even more complex. Some of the variety that Darwin observed in the finches of the Galapagos may have resulted from a specific aspect of evolving awareness (language) and not from drifting genetics.

Implications

Introducing the concept of the meme, as equal in influence to the gene in natural selection, highlights awareness as more than just a by product of living forms. Seeing awareness as an essential element for life and adding it to our criteria for defining life, allows us to appreciate its role in evolution and to begin to trace the evolutionary history of awareness as equal in importance to genetic evolution. Viewed in this light, the evolutionary impetus for continual advances in awareness explains why these advances have been naturally selected. Even the simplest of living forms must interact with its environment if it is to produce energy and reproduce.

To the three established defining criteria for life:

1. *The ability to produce energy,*
2. *Having a membrane that separates it from its environment,*
3. *The ability to reproduce,*

We should add:

4. *The ability to sense and interact with the environment, (awareness)*

CHAPTER EIGHT

Chapter Contents

CHAPTER EIGHT

AWARENESS AS A STATE OF ENERGY

Awareness is a state of energy, not a state of matter and, fortunately, the ability to generate energy is an essential quality for life. The complex arrangements of matter that make up living forms create energy using chemical processes. Living energy existed before the first proto genes divided and prompted their chemical repair to form two new forms consisting of an energy by-product that made the replication process self-sustaining. The raw materials of life involved in these self sustaining, self powered, chemical reactions are dependent on other basic chemical elements being incorporated into the process and used for fuel. Life is a slow consumptive chemical reaction creating energy used in different ways by different cells. Some use the energy to contract, others to create an electrical charge, and others to act as pathways and conduits. Disassociated, these cells survive but seem insignificant, but joined in symbiotic arrangements in complex life forms, their energy interactions add mobility as legs, fins, and wings, and awareness as eyes, ears, and brains.

We are the ooze that squeezed through the mesh of environmental testing that brought with it the energy producing processes that sustain us. This same energy is the energy that not only powers and sustains all living things, but also powers awareness.

Awareness as an unfolding dimension

The energy powering *living matter* is evidenced in a living form's ability to grow, move, and generate heat. The energy powering *awareness* is evidenced in a living form's ability to respond to rapidly changing situations. Both use the same basic energy source, (energy produced within individual cells). Single celled life produces only small amounts of living energy and has limited awareness. Multi-cellular life is able to compound the energy produced by their many cells. They do this by creating groups of cooperating specialized cells, by storing their living energy for access on demand, and by selectively directing energy to various functions based on priorities. The functional complexity of multi-celled energy systems evolved in concert with advances in internal and external awareness. Complex neurological networks developed in concert with function specializations, followed closely by priorities for internal energy sharing. When we dissect and observe this amazing maze of interconnected cellular functions, we see something familiar. We see a parallel in the interconnected parts, development and function of complex machines created by man and are seduced into drawing the conclusion that there must therefore, be a creator directing the development and function of complex living things. The evidence however, is to the contrary. Only if we define all of nature, the entire universe, as the creator, can we explain the developmental mechanisms we observe in nature.

Buddha spent much of his life, searching for a creator and a purpose through spiritual experimentation only to come to the same conclusions we have reached in our scientific investigations; nature is all-inclusive and includes our own awareness as an integral part of the whole. Each special niche in the environment, even if only slightly suitable, will direct life into a form that fits into that niche. The process by which nature assures the compatibility of each life form with its environment is natural selection and *awareness* accompanies each selection as an essential element.

Sense and react is the simplest evidence for aware-energy and can be observed in bacteria, protoctistae, fungi, plants and animals but the energy of awareness does far more than simply supplement motive abilities. Increasing levels of naturally selected *aware energy* have been selected as advantageous to survival with such proclivity that awareness has begun to influence the process of natural selection itself. Just as the introduction of sex, (the joining and exchange of genetic material), introduced death as an inevitable consequence, the introduction of choice, introduced a new natural selection directive as an inevitable consequence. Choice is created when awareness advances from simple reactive responses to complex synaptic retentions that offer optional responses. Choice is the ability of a life form to select between nuanced synaptic retentions and make a decision appropriate to the moment. Instant and appropriate reactivity improves the odds of survivability and procreation, but with its advent, aware energy began to alter the environment to fit life instead of the environment always altering life to fit existing conditions.

Evolution has taken place slowly with several significant intermediate steps. Powered only by natural selection's preference for survival and procreation living forms improved their advantage by taking these steps.

- single cells develop a well-formed nucleus
- individual cells incorporate mitochondria as partners in the replication process
- cells join in symbiotic colonies
- cells develop gene sharing
- life became multi-cellular

As each of these plateaus was reached, an accompanying advance in awareness took place, and the simple sensitivities of the first ill-formed cells became the advanced awareness of animals. Advances in synaptic capacity and sensory capabilities paralleled the hierarchical development of physical forms, and is still sustained by the energy producing chemistry of the cell but it is now an efficient cooperative multi-cellular process creating enough surplus energy to sustain advanced levels of awareness.

Thinking takes energy, a lot of energy. Powering the synaptic process and the activity of complex sensory organs uses the surplus electrochemical energy of living forms to create a second arena for natural selection; *adaptive awareness.*

Awareness as a natural occurrence

It is extremely doubtful that we will ever discover purpose among the patterns of the universe. The hints of purpose we see in the joining of elementary particles to become atoms and atoms to become molecules and molecules to become matter and matter to produce life are self-imposed. There are no sign posts along the path of discovery that tell us why things are the way they are. The only signposts are those that tell us, how the joining takes place. Our self-awareness is the product of these same processes and is at the root of our confusion. In the past, we viewed the universe from the perspective of the first person singular, and for most of our aware history, we accumulated information about the world around us, and the stars above us, using only our sensory organs and explained the universe using our imagination. Creation myths abound and sacred texts are full of revealed explanations.

With the advent of tools to augment our senses and the advent of the scientific method, we began to accumulate our information through group revelations and cooperative examinations, and our perspective changed to those of third person plurals but nature has created our most basic perspectives and we cannot escape our self-awareness. As a result, we continue to insert narrow first person pronouns into our discoveries. The only way to avoid interjecting more fantasy into our observations is to broaden our perspectives by making the intrusive personal pronouns, "us" and "we", meaning *humankind*, mean all living things. By including ourselves as part of the natural mix around us we avoid the self aware mirror that focus our attention back on ourselves and on unanswerable "Why" questions.

"How", is all that Nature is going offer as an explanation for mankind's advanced awareness and "How" is all that Nature is going to give us to explain awareness as common to all life. Our contemporary group revelations are gleaned from discovery and verification and are essential in sustaining the meme pools that support our advanced civilizations. The individual revelations of the past persist but provide few verifiable answers and, other than well-developed sets of moral principles, have little practical application in modern societies. Awareness is a natural occurrence brought about by the natural selection of attributes leading to aware states, not from divine directives leading to some end goal.

Awareness is a natural occurrence and understanding the mechanics of natural selection is our best hope of understanding how awareness came about.

Concentrations of energy that form and sustain awareness

Advanced awareness is the product of complex neurological systems. The energy that powers those systems is produced in the bodies hosting them.

A horse is acutely aware of its surroundings and has its awareness powered by the same energy that allows it to run. It grazes on grass, digests it to absorb essential nutrients and distributes the nutrients through its body where cells convert sugars to energy. The stored energy in the grass is sunlight converted to carbohydrates by photosynthesis. The horse fuels itself with the energy stored in the grass to produce body heat, internal functions, mobility and *awareness*. Through this process, the energy of the sun is concentrated and transferred between living forms to become the energy powering the awakened state of the horse with the horse itself being the product of billions of years of slow evolution powered by sunlight.

The scenario above, describing the reality of the sun's relationship to the horse, sounds more like a Greek myth describing a creature of the sun that one can ride, but with these kinds of miracles around us, who needs myths? The answer is nearly everyone. Most humans prefer myths because simple stories are easier to understand than nature. Even with proof that living things were collecting and concentrating solar energy for millions of years and storing it as huge deposits of oils and gasses trapped in sediments, most humans prefer not to have nature's processes explained. "Pay at the pump with a credit card" is proof enough.

In spite of humanities' reluctance to understand and appreciate nature's miracles, the process is ongoing and energy from the sun is being gathered and concentrated throughout the earth's biosphere however, naturally stored solar energy is being withdrawn and used in new ways as man makes choices that supersede natural processes.

The power of awareness as a creative force

Awareness is a state of energy. *Active* states of aware energy decide between options offered up by nature and *advanced* states of awareness go one-step further and create new options, options that nature has never offered. A bird selecting a hole in a tree as a place to raise its young is choosing between options offered by nature. A bird building a nest is using its advanced awareness to create an option not previously offered by nature. The ability to build a nest reflects the evolution of awareness, the evolution of aware energy into memes for collecting and arranging twigs and leaves that, over generations, collect into a meme pool, (how to build a nest). The awareness of nest building birds now contains a new option, a new 'natural' option with genetic support but without genetic guidance. Awareness has usurped environmental selection by being creative without using a gene pool.

A spider building a web, a turtle digging a hole for its eggs, and man building a city are all examples of awareness stepping around genetic natural selection. The energy of awareness is an extremely powerful force based on choice. The power of awareness is expressed every time a choice is made. By making a choice, natural selection is displaced, (but only temporarily). Natural selection, even when suspended, ultimately makes all final decisions.

About three billion years ago, genetic selection produced prokaryotic bacteria so successfully mated to a methane environment that they multiplied in overabundance and poisoned the atmosphere and themselves. The genetic disaster put life on hold until natural selection re-adjusted the balance between living forms and the environment. Natural selection had only one option, adapt living forms to the new environment using the same methods that produced the imbalance, mitosis. Over time, genetic drift produced new single celled creatures capable of using the oxygen rich atmosphere to their advantage, and from the higher energy levels of oxygen breathing life, came multi cellular forms and advancing levels of awareness. Awareness was simple at first, but by collaborating with successful genes avoided de-selection and increased in scope, span and power. Now the advanced state of man's awareness has reached such an extreme that the overall state of natural selection has once again become unnatural and maladjusted. The power of man to create new options has overridden natural selection's balance, and like the imbalance created by the great proliferation of early single celled life altering the atmosphere, the power of man's advanced state of awareness now taps into the energy of matter itself, and manipulates genetic selections to its own purposes. The environment is no longer the sole pattern against which life tests its appropriateness. The awareness of man now provides alternate prime directives similar to the time when the composition of the atmosphere was changed, the results of the expressed power of these new choices challenging natural selection, awaits resolution.

The impact of awareness on evolution

Many animal species have evolved advanced states of awareness, but only man has taken awareness to an extreme cognitive level. Species with an evolved active awareness influence the biosphere directly by making choices that alter the environment. Species with only a passive awareness affect the environment only indirectly by their presence. Primitive species with only a primitive awareness are difficult to distinguish from the environment itself. The internal processes of primitive life are organic, but their internal chemical interactions closely mirror the inorganic reactions in the environment and a distinct line between the evolution of the environment and the advent of life is difficult to establish. Life is a natural extension of the environment and evolves in concert with the environment until an active state of awareness establishes a separate course. With the advent of an evolved active state of awareness, life became self-directive and could ignore environmental demands.

Passive aware states adjust the environment by their presence but not by assessing and adjusting their surroundings. The interjection of aware choice into the arena of physical adjustments alters the path of natural selection. For most actively aware life forms, the impact of their choices on the environment is a minimal, but in man the natural selection process directed by nature is being usurped by the power of his cognitive awareness.

The impact of mankind's awareness was minimal until he began to gather in social organizations larger than tribes. For most of human history, the environment dictated man's adaptations and his environmental impact was limited, but his influence increased as he used language skills to developed larger social organizations. Villages became cities and cities became empires as language facilitated his cooperative efforts, and with civilizations holding natural selection at bay, human populations began to expand exponentially, but natural selection intervened often.

Man has challenged nature many times. He disrupted a natural balance with the world of microbes by creating cities and unsanitary conditions new to nature, and suffered many plagues. He disrupted natural landscapes and suffered droughts and floods, and has created ideological constructs and boundaries and exterminated himself' in great numbers defending territory and beliefs, but the real impact of man's cognitive awareness began with the advent of the investigatory method called science. The result of a product of the environment investigating, the environment from which it came, is yet to be determined.

Awareness and memetics as an arithmetic progression of coded information

About four hundred years ago, myth began to give way to observation and a new perspective emerged inspiring confidence that nature could be understood if we looked more closely and shared our observations. By applying mathematical principles to careful observations, we set in motion a self-perpetuating process of data accumulation that is still expanding exponentially. Understanding how nature works and is organized has allowed us to envision and build tools to peer even deeper into nature's secrets. At first, we used the deductive conclusions of philosophers and our naturally selected and ingrained logic as our guides to mine the content of memes hidden in our complex languages. The deductive conclusion that, "if we continue to divide a substance into smaller and smaller halves we will eventually arrive at an indivisible substance," led us on a quest that has produced the microscope, the electron microscope and the Large Hadron Collider. The chain of tools used in this quest was the sequential of discoveries made by each preceding tool. The same chain of discovery has led us to our advanced understanding of astronomical principles, biological principles and medicine. The trillions of memes and concepts that now make up man's powerful meme pool exceed any individual's synaptic capacity and exist only because of other inventions produced in concert with the chains of discovery described above. Among these are, written language, mass printing techniques, formal education methods, electronic communications, libraries, computers, and the web. The essential meme pools that sustain humankind, are artificial and external, and exist primarily as information maintained by tools invented to augment man's synaptic, aware capacity.

Man's social networks have adapted themselves to access, use, and sustain these essential supplementary information centers in a process of natural selection much like genetic adaptation. In an even more generalized perspective, the powerful meme pools of memetic information being collected and used by humankind are only a hierarchy of codes, directing concentrations of aware energy.

Genetic codes spell out and direct the ordering of organic molecules into living forms. Memetic codes spell out and direct the ordering of synaptic pathways into concepts and perspectives. Language codes spell out and direct aware energy into meme pools creating an aware energy reserve. Enhanced communication and transmission abilities augment and enhance both language codes and memetic codes. Binary computer codes concentrate aware energy and advance man's investigative and innovative capabilities.

Each level of code is an expansion and interpretation of its parent code and as a result is transitory. Only the original code contains its own proof. The extreme power of Man's concentrated aware energy rests on a series of fragile constructs, all of which are code dependent. Assessing the vulnerability of these essential codes and devising ways to protect them should be one of our highest priorities. In our digitally dependent age, if the electrical grid fails, all the books go dark, and we are unprepared to survive using the skills and simple tools of our ancestors.

CHAPTER NINE

Chapter Contents

CHAPTER NINE

COMMUNITIES OF AWARENESS

Evidence for communal awareness

Advancements in awareness through the natural selection of sensory organs and neurological interpretive centers have been a priority for natural selection from the beginning. The grouping of cells, the grouping of plants and animals and the creation of pools of awareness for survival and reproduction, has also been a priority created by the survival benefits of grouping and increased complexity.

Before the living cell became perfected, communities were forming and being selected as advantageous. Precursors to multi cellular forms still exist as loosely joined interdependent small-specialized individual organisms and are nearly indistinguishable in form or function from multi cellular life. Both colonies of organisms and more familiar multi cellular organisms are examples of living communities of individual cells that communicate and cooperate to produce a survival advantage. The next hierarchy of groupings to add survival advantages was the gathering of multi cellular life to form packs, pods, flocks, herds and tribes.

The cohesiveness of every grouping depends upon a continuing mutually beneficial interdependence and some level of organization. The positioning of primitive cells with flagella to gather nutrients adjacent to cells capable of fixing minerals for anchoring is an example. To control breathing, blood flow and digestion in more complex groupings sophisticated internal chemical communications networks were also required. The simple transfer of information between internal organs to coordinate their activities and to accommodate immediate needs occurs without continual oversight by an active awareness, but benefits by an occasional intervention.

The advantage of community continues beyond the inner workings of multi cellular forms. Complex communities of awareness, like communities of cells and organs, persist because of their survival advantages. Organizations of multi cellular organisms collect as communities of awareness because of the advantages they create for individuals comprising the group. Community awareness is created when individual information is shared. Group pools of awareness can be instinctive, emotional or intellectual.

Pools of awareness provide various levels of survival advantage and can be both of benefit to the group and a threat to individuals outside the group. In extreme cases, threats to others can results in the extinction of other groups of the same species.

Community as a stimulus for advanced awareness

The partnership of genetic/memetic evolution has encouraged communal living and the formation of shared pools of communal information. The bonding of mates for life, the gathering of schools and pods, the aggregation of herds, coveys, and swarms for sustenance and survival, are only a few examples. Group behavior memes supplement genetic instructions for grouping by adding learned patterns for detecting food and predators, better group defenses, and other valuable bits of shared communal awareness. The gene cannot provide these advantages but does provide the ability to acquire them.

The body of behavioral information sustaining a group is a shared form of individual response information but has a coordinated survival value beyond that of an individual. Both the natural combining methods that create atoms molecules and stars, and life's natural adaptive development methods trend toward the more complex. The addition of communal awareness is a prime example of naturally added complexity.

To be of social value, information transfer must take place, an information pool established and the pool sustained by all or most group members. In species without a written or alternate way to store information, a continual teaching / learning process is required.

The genetic advantages of community have primitive origins. Beginning with single celled life gathering into hummocks, the genetic advantages of community have continued by pairing with neurological advances in multi cellular life, becoming instincts, (gemes), and with the advent of active awareness in mammals being supplemented by (memes).

Group instincts, (gemes), are essential for many survival behaviors including the protection and nurturing of slow developing young. In these species, the young are defenseless and require constant care and supervision for years. In many cases, the disadvantages of slow development have been overcome by the advantages of strong family and tribal support systems. Large brains and well-developed languages are common characteristics of mammals that require long nurturing periods.

Birds also exhibit strong family bonds along with cooperative behavior in larger groups. Many birds exhibit well developed traits of advanced adaptive individual awareness and well-developed complex languages, but, beyond short egg incubation periods and short nurturing periods, most birds can survive as individuals early in life. In many species of birds, the acquired need to migrate may have been the impetus for the development of communal awareness and advanced aware states.

Bird and mammalian brains share many common characteristics exhibited by an advanced aware state. The much smaller brains of birds appear almost as capable as larger mammalian brains, and have a much longer evolutionary history, evolving as a continuation of a several hundred million year long evolutionary history as dinosaurs. As a result, the brains of birds contain many more neurons per gram than mammalian brains and are more efficient.

Other species, within the phyla chordate, also exhibit advanced levels of communal awareness but appear to have had the required individual characteristics selected for different reasons. The only other animals that have become terrestrial, (arthropods), include many species that display advanced forms of communal awareness. Some insect migrations cover such large distances and require two or more generations to complete. Understanding the method of information transfer required for this feat is beyond our reach and is another example of the strange abilities exhibited by creatures that undergo a metamorphosis. The complex communal behavior of other insects, like ants and termites, may be evidence for the existence of a chemically based communal intelligence that we have yet to understand and appreciate. Unlike animals that concentrate large amounts of neural capacity in the individual, the arthropods use extremely effective external methods to link limited individual capacities to produce an entirely different type of advanced awareness, with individuals acting like mobile neurons.

In man, articulate tongues, opposable thumbs, upright walking and an omnivore diet, were enough to keep humans on the selected list long enough for them to add language, fire, and tools, to their naturally selected advantages. With the addition of self-created assets, mankind has become the most preferred of all naturally selected life forms and has further exploited his advantage by becoming a symbolic thinker. Man is the prime example of community influencing the development of advanced awareness, From man's early cave paintings, he developed written language and the ability to store community information beyond individual life spans and communicate through time. Writing has taken human communal awareness to a much higher level. The potential for species living in communities to produce levels of awareness beyond the limitations of individual retentions has created a communal awareness in many species.

Social awareness and empathy

To provide a survival advantage, the awareness of ones surroundings must include the ability to recognize the difference between life forms that pose a threat and those that do not. Awareness must also include the ability to recognize like kind, even when the differences are subtle. This naturally selected ability exists at all levels of awareness and is essential for the perpetuation of all species. Dogs recognize other canines as members of their species even when size and appearance varies greatly, as in the difference between a Great Dane and a Chi Wawa. In contrast, many birds recognize that other birds with an almost identical appearance are not members of their species.

The ability to recognize like kind and the sex of a potential mate has developed as a meme in concert with genetically sensory abilities and implanted responses. At primitive levels of awareness, the recognition and selection process for a mate is primarily biomechanical. At more advanced levels of awareness, memes begin to play a larger role in the final selection process.

The ability to recognize members of one's species, members of one's pod, flock, school, herd, or a tribe, or members of one's family, adds survival value through both inclusion and exclusion. Being able to discern small nuances of individuals allows the development of smaller social networks and specialized groups sharing common characteristics. Both genes and memes play a part in this ability.

Selective recognition adds an additional level for natural selection to coordinate genetic testing. To support selective recognition, larger and larger portions of neurological centers have been genetically selected to support detailed synaptic recognition templates. Memetic and genetic recognition methods are now thoroughly intertwined. When learned memetic recognition refines a choice made by an instinctual gemetic response, both help avoid a genetic rejection of incompatible DNA.

Memetic genetic recognition evokes emotional responses that are; protective, aggressive, and empathetic. Being able to empathize with a member of one's group or species has survival value and is a preferred genetic and memetic selection.

Empathy; the ability to identify and sympathize, has become so ingrained that it often extends beyond family and beyond like kind groups to other species and can be used as an indicator of cognizant self awareness. Many animals show empathy for the unattended young of another species and appear to sympathize with animals of other species in distress. The ability to appreciate emotional responses in individuals of different kinds reflects the ability to projected oneself into another's situations in an appreciation of the others awareness. Empathy and compassion are words descriptive of a basic universally selected set of enabling emotions. In the competitive arena of life, where the elimination of competitors is also emotionally driven, empathy may be the essential counterpoint that allows a sustainable biosphere to develop and persist.

If we restrict our definition of empathy to emotional identification with members within a group, we can begin to appreciate the strength of these momentary bonds. A female Gibbon observing another Gibbon mother's tenderness in caring for her young understands the emotions involved both instinctively and from memory, especially if she herself has raised an offspring. The emotions of the mother gibbon are recreated in the observer, vicariously. Both mothers connect by a common momentary emotion. The almost telepathic message is, "I am you and you are me and we understand each other."

The shared emotions between a mother and the child are similar and are a form of empathy. We have labeled many of these emotions instincts, but empathy requires more than simple knee jerk passive awareness. Empathy requires active awareness in order to assess the immediate situation and at least enough cognitive awareness to recognize and project an image of self into the situation. Empathetic emotions result from projecting one's self image into the action, and when several individuals participate in this activity, group dynamics override individuality.

Social cognition and language

Language has no value beyond the exchange of its content. Language uses many methods for transmission and reception. These can be chemical, postures or gestures, symbolic vibrations, visual symbolism or digital pulses. The value of language is its ability to facilitate the formation of social groups that, in turn, adds a favored natural selection status to individuals within the group. Language adds survival advantages to swarms, schools, pods, herds, flocks, and tribes. In man, the nearly total symbolic representation of the environment in written and digital form allows an exchange of information beyond individual conversations and far beyond the content capacity of individuals.

The verbal and visual languages of man probably developed together. Using a scratched symbol to represent a spoken name, rather than relying on a poorly drawn pictorial representation, became a preferred method because it promotes accuracy by relying on man's universal ability to modulate and interpret sounds rather than a universal ability to produce and interpret difficult artistic forms. The Korean alphabet is probably the best example of this advantage. The Korean alphabet was intentionally created to depict spoken sounds and to establish a new secret set of sound symbols replacing their established pictorial representations. The intent of the new alphabet was to make their writing unintelligible to an oppressive occupying force, not to invent a new language. The language didn't change, only the symbols for its sounds changed and the logic of combining consonants and vowels in easily recognizable syllables makes the Korean alphabet unique and easy to read.

When one reads a written language, the symbols on the page are recognized as sound and are spoken in the mind as a synaptic transfer from the visual cortex to synaptic retention centers associated with the interpretation of sounds. The internal interpretation process must be learned and is the foundation of all formal education systems. To learn to read, one has to develop an internal visual/auditory interpreter. The simpler the visual symbolism the faster and more accurate the internal translation becomes, and the easier the internal interpretive skill can be developed.

Social cognition is almost entirely dependent upon language. Spoken language provides real time adaptability. Written language transcends time and allows information to be stored and accessed later, as needed. The sharing of information evolved to include, complex scientific information, mathematic, and philosophical dissertations, and is the basis for the largest of social cognitive groups (civilization). Recognizing the value of social cognition and the value of the languages that make civilization possible highlights the importance of educational enterprises that sustain both. It also highlights misdirected enterprises seeking isolation and simplicity.

CHAPTER TEN

Chapter Contents

CHAPTER TEN

SELF-AWARENESS

The evolution of neurological systems

The physical organs that create awareness are the most marvelous of adaptive natural selections. Initially created by genetic drift, neurological systems developed additional complexity as they were selected for their ability to make rapid adjustments to fast changing environments. Progressive selections for increased awareness became the awakening of living matter, a shift favoring the genetic codes for *sense and respond* abilities. Over time, these codes became dominant. One could argue that there is inherent in the basic structure of matter, a hidden pattern placed there to promote directives in evolution, but if these exist, they are hidden and beyond our investigative abilities. To support such purposeful assumptions we invent scenarios without any observational verification and turn our back on what nature is actually telling us about her creative processes.

We continue demanding to know "Why" while nature repeatedly responds, "I can only tell you how." When we stop inventing answers and read the book of "How" that nature has

---

opened for us, we discover a process far beyond the trivial stories created by our imagination and far beyond the revelations of ancient prophets. The process is simple and yet so powerful that it is immediately both understandable and amazing. Nature's method is the miracle of matter organized and shaped by natural forces into the most creative process imaginable. If we could copy the process, all things that we could possibly invent would be created without our intervention. We could step back, rest in our creative efforts, and watch future inventions unfold before our eyes. Unfortunately, we cannot step back because we are an integral part of the unfolding creative process we are observing. We are the result of awareness expanding through genetic selections because of its favorable survival value. We are the most awakened state of matter. We are nature's latest and most extravagant test of advanced awareness, and as long as our advancing awareness continues to prove worthy, the genetic arrangements that produced our amazing synaptic capacities, will continue.

Our awareness has no physical attributes, but the complex neurological systems that produce it are open to inspection. Nerve cells are similar in Arthropods and Chordates. The typical nerve cell,(neuron), is composed of a cell body, a nucleus, small branching protrusions called dendrites, and on some cells one or two lengthy protrusions called axons. Nerve cells can transmit and receive signals from adjacent neurons through their dendrites or can receive and transmit signals over feet or even a few meters through long specialized myelin sheathed, (insulated), axons. Arthropods and other invertebrates depend more on reflexive networks while most vertebrates have developed and depend more upon elaborate living computers, bundles of millions and in the case of man, 100 billion neurons that exist in close proximity, and are interconnected by tiny dendrites and long axons. The flatworm's entire body is encased in a network of interconnected neurons. The lobster has a somewhat more developed reflexive system with a simple grouping of reactive neural cells called ganglia forming a simple brain. All animal brains have the same structural parts, evolved from the ganglia networks of invertebrates, and all retain these primitive structures as essential parts of the more complex brains that direct their activities. The electrochemical process used by neurons to generate and pass along signals is the same for all aware life forms. All life uses the same basic equipment and the same energy systems to generate awareness and depend upon the same natural creative process to produce their neurological systems. An examination of the human brain reveals all of its evolutionary precedents still in tact and operational, but substantially modified and overlain by billions of additional neurons for information processing and storage.

We have studied our own brain in detail. As a result, we know a great deal about its structure and function. We have also made in depth comparative studies of the brains and neurological systems of other animals and can identify the genetic trace leading from the simple to the complex and from the earliest of sensory capabilities to the cognitive awareness of man. For those seeking more details, libraries are filled with volumes of detailed accounts of neurological structures and their functions. The most complex systems that has ever existed, the living arrangements of multi cellular neuropathic structures that produce awareness, have been created by the simples of all processes, *natural selection.*

Internal sensors and passive self-awareness

The ability to sense ones own internal condition has evolved in parallel with life's ability to sense its external surroundings. Complex single celled life responds to disruptions in internal conditions using very simple monitors. Multi cellular life uses similar cellular monitors located strategically within key organs that are connected to primitive parts of the brain through chemical secretions and by long networks of nerve cells. These essential monitoring systems are common to all living forms from the primitive to the complex and the number and types of these internal sensory systems in other phyla and species is being categorized and studied in detail. Understanding the commonality and diversity of internal monitoring systems in many life forms will become even more important as we advance our ability to manipulate genetic information.

In a single celled organism, a prompt by an essential enzyme can trigger a reactive response. In advanced multi cellular creatures, many more systems are monitored, and complex chemical and electrical systems prompt alerts. Self-awareness at the level of internal monitoring systems is a naturally selected sensory/synaptic capacity carried forward genetically as an essential survival attribute. Self-corrective body repairs, such as the production of antibodies, swelling, the congregation of white blood cells, etc. are also an essential part of this self-aware regulating system. In most cases, even in advanced life forms, the intervention of active awareness is not needed to activate these corrective responses, and higher levels of the brain are not involved. Awareness at this level is passive, but remains continually on alert.

Levels of self-awareness

Passive self-aware responses to an internal alert are reflexive; scratch the itch, avoid stepping on the wounded paw, etc. Active self-aware responses to internal alerts are directed; climb the hill to find grass, lay in the sun to relieve an aching joint etc. Active awareness also produces the beginnings of recognition of self as separate and distinct from others. Levels of self-recognition parallel levels of neural development and is more advanced in creatures with levels of active awareness approaching a cognitive level.

Cognitive self- awareness is the ability to recognize self as a separate entity and evaluate one's self image as if it were a separate like kind and make behavioral adjustments based on the evaluation. This, almost schizophrenic ability is most prevalent in mankind but is evident in some primates and marine mammals, and possibly by a few advanced arthropods as they respond to their reflected image with what appears to be self recognition. This level of self-awareness has most likely developed from the importance of being a part of a group where relationships within the group require self-evaluation to form appropriate responses to changing group dynamics. Cognitive self-awareness is suppressed during periods of intense focus on functional or directed mental activities. It only becomes evident during more restful moments. Our advanced mammalian cousin's may know nothing of yoga but seem to practice it in a primitive form none the less, and one has to wonder when observing pensive moments in more primitive life forms if their awareness is not inwardly focused.

Cognitive self-awareness

Advanced cognitive self-awareness in Man has evolved through natural selection to make us the most self-aware life form on the planet. Our intense ability to think of ourselves as separate from our own thoughts is the product of the peculiar way in which we are hard wired. This ability to recognize our own awareness has also created our self-aggrandizing self-images. From our point of view, when we observe our surroundings, we are observing nature in its entirety, and operating efficiently in the totality of all existence. Our ability to manipulate nature and to create cities and other great works reinforces our inflated self-image, but in fact, we are observing only a sliver of the totality around us and acting out our lives on a very small stage.

The cognitive self-awareness of man also contains the realization that one's awareness is subject to mandatory termination when one's physical support systems reach their longevity limits. Being aware that one's awareness will end is the root of religious beliefs, ritualistic sacrifice, and our desperate attempts to extend life by any means possible.

Any cognitive awareness contemplating its own inevitable demise creates a closed loop that is incompatible with other thought processes. The end of ones awareness is incomprehensible because using awareness to explain a permanent state of being unaware is impossible. The result is fear, the kind of fear that would exists if one were to stand at the entrance to a dark cave without any information as to what is inside, knowing you will have to enter alone, and that there is no direct evidence that anyone has ever come out. This fear is the impetus for myths and religions and is used often by those seeking power. Curiously, the dark cave analogy above is the inverse of the cave analogy used by Plato to explain human awareness. In our analogy, mankind is standing outside the cave unable to see into the dark. In Plato's analogy, mankind is standing inside blinded by the light from outside.

Knowing that the immortal method of single cell division, mitosis, could not lead to self awareness and that the natural selection of sexual reproduction and its inevitable consequence, death, were essential to the emergence of self awareness, should give us some comfort in our dilemma. Without the introduction of meiosis and individually separated lives ongoing as sequential sparks of awareness, the most advanced forms of life would probably still be pond scum. Personally, I accept the death sentence in exchange for the opportunity to experience life at a cognitive level and explore the universe for even a short period.

The evolution of living forms and of awareness is possible only if the new replaces the old. Nature does not hoard anything. Instead of whining about our terminal condition, we might try enjoying the ride, and do our best to leave behind, something of value.

Self-awareness, as with all naturally selected attributes, is a mix of genetics and memetics, a mix that can be positive, neutral or detrimental. The positive advantages of self-awareness outweigh its inherent faults. The selection and elimination process continues and we are only partially in control. Our sense of self is reflected in our languages, our self-esteem, our self-pity, our self-control, and our self-motivation, and attests to its importance in our lives.

The Gödel loop

The paradox of self-awareness can best be illustrated by our boldest attempts to develop mathematical systems and proofs. Mathematical symbols and formulae mimic our innate logical processes but eliminate aware content and retain only relationship, form and quantity. Among the many logical systems that have been created to explain complex mathematical relationships are Euclid's Elements explaining the geometric relationships of pure form, Newton's Calculus to explain the relationships of changing variables and Whitehead and Russell's *Principia Mathematica* to codify and explain number theory. What Whitehead and Russell claimed to have created was a logical structure, based on accepted axioms that could be used to explain and prove all numerical relationships. They created new symbols to avoid unintended or hidden content and carefully constructed a logical web of proofs.

Then came... Gödel!

As a young Austrian logician, Gödel was uncomfortable with the new symbolism and with the self-generation of proofs proving proofs. After analyzing Whitehead and Russell's work he demonstrated that the complex formulae of the *Principia* could be represented equally well by very long whole numbers, and when represented in this way, the inherent proof claimed by Whitehead and Russell, did not exist. Their formulae could be correct but did not contain a self-proving structure. His analysis showed that symbolizing a symbolic structure results in a strange loop that eliminates certainty.

Self-awareness is the symbolization of another symbolic structure (basic awareness), and results in a Gödel loop. If we try to prove our own awareness using our self-awareness, we end up with an assumption that may be correct, "Cogito Ergo Sum", (I think therefore I am), but we can never prove it. We are forced to rely on other aware forms around us to substantiate our own awareness and can never reach certainty. The result of this disconnect is the root cause of our susceptibility to myths promising a true realization of self through religious and patriotic purposes and allows the unscrupulous and power hungry to manipulate the emotions that accompany our inherent uncertainties. The fearful and insecure are easy to lead.

Finding ourselves in the aware reflections of other's is the emotional base for family, tribe, team, company, country, congregation etc. It also results in cults, gangs, and radical organizations that are seeking a way out of life's complexities through isolation and limited perspectives. The uncertainties created by cognitive self-awareness enhance the need to belong, promotes the creation of armies, and the misdirection of individual perspectives. It is at the root of most human conflicts but remains essential to human existence. We are trapped in a strange loop that requires acceptance but creates avoidance and is the reason we have created gods and religions. There may be no escape from this dilemma, but we can minimize its negative impact by recognizing its roots and trusting in our observations.

CHAPTER ELEVEN

Chapter Contents

CHAPTER ELEVEN

LANGUAGE AND REASON

The evolution of language

If we define language as the transfer of information between individual life forms, we make life and language synonymous. Life exists as a special arrangement of organic elements that replicate by transferring information. The information is transferred in a language of carefully arranged amino acids and is passed sequentially when a cell divides to produce two sister cells. In a very real sense, DNA is the language of life and was probably the initiator of life itself. First, there was the word, then there was life, and they will both continue to exist only if their essential partnership continues. Without a living form to speak the language of DNA there is chemical silence, the process stops, and all life ends. Likewise, living forms must continue to speak by sharing DNA or there will be no replication and the silent generation becomes the last generation. The most primitive of living forms communicate chemically. Prokaryotic cells occasionally pass DNA messages through their outer membranes when forced into close contact and bacteria have been observed excreting pieces of DNA that are subsequently ingested by other bacteria. These earliest methods of transferring genetic information are languages, a primitive form of language, but nonetheless a language.

Language as an attribute of life has its own history of natural selection and is as diverse as life itself. Language is at the core of life and has been a partner in every step of all natural selections. Genetic communication supplemented by chemical interactions was the primary language used during life's early years and has added many dimensions as its survival value has increased.

In multi-cellular life, language is selected as a survival attribute frequently. Complex languages are the product of group awareness, have their own survival advantages, and living forms remain isolated if they are without an ability to communicate. The ability to communicate can take any form that is compatible with the level of development of sensory organs and neurological capacity. It can be as simple as the flick of a fish's fin, the pheromone trails of insects, or as complex as the dance of a honeybee, the flickering florescence of deep-sea creatures, or the articulations of man.

Just as the genome generates a new living form during reproduction, language carries memes to other living forms during receptive periods of awareness. Just as genetic modifications of physical characteristics are tested against the environment, and when successful passed on, memetic modifications in behavior are also tested, and when useful passed on. Both successful genetic traits and useful memetic traits become a permanent part of a species survival repertoire.

The development of language has accelerated the process of memetic natural selection just as the advent of multi-cellular life accelerated the process of genetic natural selection. Both created better methods of producing variant forms. When more genetic combinations are available for testing, survival and mating advantages improve. When more variant memetic combinations are available for testing, valuable survival responses will be ingrained into social group memetic pools more often. The natural selection of a meme pool is no different from the natural selection of a gene pool. A healthy reference pool helps to insure the future of those with access to its values. Advanced language increases a meme pool's value exponentially and accelerates meme transfer. Both gene pools and meme pools morph over time through drift and in response to environmental changes. Natural selection favors increases in the complexity of language just as it favors an increase in the complexity of physical form. Both add survival advantages. The advanced languages of man are the result of natural selection just as humankind is the result of the natural selection of physical forms and advanced awareness.

Reason is an attribute added by an advanced language capability. The ability to observe and recognize individuals, objects, arrangements, situations, conditions and movements, is common to most animals possessing an active awareness. Assigning these objects and situations an auditory symbol, is also a common trait among animals, but complex symbolism, capable of carrying conceptual content, is unique to Man and is the root of Reason.

Cognitive thought uses extremely complex neurological pathways to interpret and analyze situational conditions before choosing a response. Reason introduces a new tool into this analytical process. Symbolic manipulations using the naturally selected logical processes embedded in language allows the analysis of more complex situations. Advanced language capability is an evolutionary jump comparable to the advent of meiosis.

Early forms of symbolic representations, such as cave drawings, give evidence that our advanced languages have evolved from early forms selectively. The functionality of any modern language is the result of it being matched to reality through trial and error. Grammatical arrangements are the embedded logic of successful natural selection and the fullness of a language's vocabulary is the result of usage. Like awareness and physical forms, languages are in a constant process of evolution. Reason is the ability to use language symbols and natural symbolic arrangements to analyze complex situations.

As a language develops complexity, it begins to operate below the surface of consciousness. Like a net, an active language collects ideas, relationships, new words, and new meanings as it is dragged through both social and natural environments. The collections in the net become visible only when we pull it to the surface to converse or to examine its content. Language nets catch and hold aspects of the natural world and their relationships beyond our intentional additions. We analyze these unexpected catches by sorting through them using a process called deductive analysis. Before the invention of the scientific method, deductive reasoning was the primary method used to unravel the mysteries of nature. Evidence for deductive explorations exists in the writings of ancient philosophers, especially in the dialogues of Plato. Deductive reasoning remains a valuable tool, and when used in combination with inductive reasoning forms the foundation of man's success.

As our engaged enterprises force us toward a common language we need to preserve the content and wisdom contained in languages fading into extinction. Within endangered languages are active meme pools that cannot be accessed after a language dies and the wisdom hidden within them becomes empty words upon a page.

Symbolic thought

Cognitive processes are wound tightly around symbols and analogies as the naturally selected best method for creatures to interface with a complex reality. As a preferred trait, this method became an integral part of active awareness early and an essential part of man's cognitive awareness as he began to use a definitive language. We think in symbols, speak in symbols, write in symbols and symbolize ourselves. An example may help.

"As you sort through some clothing you come across a shirt with a logo, (symbol), on it that reminds you of a toy, (symbolic representation), you wanted to send to your grandson but had forgotten to mail. You package the toy and put your grandson's name, (written symbol that identifies your grandson), on the package along with the address, (alphanumeric symbols for other symbols representing houses, streets, states and shipping routes). You pick up the phone and call, (using numeric symbols that represent other name symbols for a business), FedEx, (more symbols), and request a pick up. They ask you for all your symbols and give you a symbol as a confirmation. FedEx processes using symbols, drives a truck with every part of the truck initially symbolized on a design drawing. They look for your grandson's house and street symbols to locate him then pass the package through a complex symbolic sorting process that finally gets it on a plane, (the product of tens of thousands of symbols), where the pilots look at symbols to fly the plane, talk to air traffic control in a symbolic language etc. etc".

Further complicating our adaptive (or maladaptive) ability to think and converse using symbols we also use analogy.

Analogy and metaphor

Many of the memetic groupings that become themes and perspectives are beyond visualization or a quick logical analysis. To compensate for our conceptual limitations we describe difficult concepts in an understandable form and simplify them further to make them suitable for language transfer. We call these simplifications analogies. We create a visualization that can be more easily explained and transfer the simpler image to others using common language along with instructions for them to extrapolate to a larger, or smaller image, to understand what we are really trying to convey.

For those who don't understand logarithmic progressions, and have never experienced an earthquake, the Richter scale for measuring earthquakes has little meaning. Without prefacing an explanation with lengthy mathematical instructions, explaining the difference between an earthquake measured at three on the Richter scale and an earthquake measured at seven, would be nearly impossible. If we use the analogy of a mouse shaking a small sapling, compared to a gorilla shaking the same sapling, however, the large difference between the three and the seven become evident. It is up to the recipient of the explanation to extrapolate from a sapling to a building. The listener may not extrapolate at all, but the idea that the Richter scale is something more than just simple arithmetic has been made evident.

We often preface our analogies with terms that announce the fact that an analogy follows, with phrases similar to,

The _____ is like a _____ or, a_____ is just a _____ inverted.

We can also interject analogies and metaphors without warning, assuming they will be recognized by the receiving individual. Many assumptive analogies and metaphors can be found in ancient writings and, unfortunately, are often interpreted as factual statements by modern readers who have become dependent upon formal prefacing announcements.

Analogies and metaphors fill our everyday conversations, our literature, religious writings and scientific works. They are the crafted tools of politicians and despots and comprise a significant portion of the content of our meme pools. Overused they become clichés, are misinterpreted as facts, and become contaminants in our communal perspectives. Used appropriately, analogies are essential to the meme transfers that sustain our morality, our scientific understandings, our sense of history and civilization. We could become much more secure in our self directed attempts to cooperate with natural selection if misused analogies and metaphors could all be "red flagged", unfortunately this seems unlikely.

Imagination, observation and logic

Increasing synaptic capacity continues to be genetically preferred as the accumulation of information supports survival. New pyramidal brain cells, each with dozens of dendrites branching out, interconnect and form optional pathways for information, retention, and transfer Being smart is a genetically preferred state and coupled with a genetic propensity for communal life, the combination has produced man. A genetic preference for larger brains has resulted in humankind. Except for birth canal limitations, this out of control combination of survival advantages may have continued even further. Approximately one hundred billion cells now make up the average human brain and many of these develop after birth as the skull expands to accommodate them. A human child is relatively helpless immediately after birth until its capacity for awareness and control develop. Young humans develop active meme retention slowly. Memories that carry forward to adulthood do not begin to form until language begins to develop at about age two. As brain cells multiply exponentially in a newborn's skull, billions of empty dendritic connections are made, forming the biological base for an advanced awareness.

As awareness develops in young humans, sensory inputs are at first randomly associated among the billions of connective options available in their brain, and use only a small proportion of the dendritic paths available. Once used a path leading from one pyramidal brain cell to the next remains active and, like a trail in the woods, becomes a preferred path for later neural transmissions and the more it is used the more it is preferred. Unused connections remain passive but available. From these many optional alternate paths, dreams, imagination and creativity are produced.

The brain's electrochemical activity seems to pulsate and can be measured as brain waves in cycles per second. Certain types of mental activities are associated with the rapidity of the pulsing. Rates between one to three cycles per second are associated with deep dreamless sleep. Rates between four and seven cycles per second are associated with dreamy, creative intuitive states. In deep sleep, all aware pathways, except those monitoring and controlling essential bodily functions, appear to be quiescent. In dreamy, creative, intuitive states, random pathway exploration using established connections and unused connections appears to take place. It is from this semi-aware state that we imagine things and situations that do not exist. Most make no sense but some curve back on themselves forming something recognizable that could or might exist. These imagined possibilities become the many works of man. By letting our synaptic capabilities randomly associate within limited parameters, we create art, music, literature, humor, tools and civilization. Our random associative creative abilities are useless, however, if they have no relevance.

Music without the disciplines of rhythm and harmony is just noise, art without composition and form is just nonsense, and literature without grammar and meaning is just gibberish. We develop the associative disciplines needed to make our random creative abilities relevant through observation, trial and error, and education. We develop and maintain relevant synaptic traces by observing our surroundings, by mimicking the behavior of other species, and through instruction. Over time, we develop a repertoire of appropriate responses that allows us to function effectively in our environment and our social settings as we continue to observe and adjust our activities to interface effectively with reality.

The essential patterns of appropriate responses created to deal with the complexity of our ever-changing surroundings, are synaptic traces that produce logical behavior and define human objectives well.

To access and operate within the arena of active reality we voluntarily increase our brain wave activity to between fifteen and thirty cycles per second, with the higher rates associated with periods of focused concentration. In humans, even simple daily activities are complicated by symbolic overlays and require priority selections.

Instead of dealing directly with nature, we modify our environment by ordering it to fit our perceived needs. The power to do this comes from our ability to think logically and from our complex and powerful language. We turn our synaptic traces into symbols that we can share and store, and twist them further to our advantage by fitting them into an order we call logic. At its root, this advanced manipulative ability is simply the effective application of rules we observe in nature when events result in predictable consequences and as we observe inclusive and exclusive groupings. We have taken advantage of our ability to observe these subtleties by learning to symbolize them, to remove their content, and then to manipulate the empty symbols to predict patterns and outcomes we can apply to other situations.

CHAPTER TWELVE

Chapter Contents

CHAPTER TWELVE

HUMAN AWARENESS

The impact of the environment on the development of human awareness

Anthropogeny is the investigative study of the origin and evolution of Man. It uses scientific methods to examine fossil and anthropological records to study the many human species that have evolved. Homo sapiens are the only extant species of the genus Homo australopithecines.

Homo erectus, and Homo sapiens Neanderthal are extinct, but our extinct relatives and we have common ancestors as evidenced by our DNA being 98.4% identical to chimpanzees. This statistic becomes even more impressive when we realize that the DNA trace, along life's family tree leads directly to our primate ancestors and then further to the tiny root of bacteria where life began. Our DNA contains 2,000,000 alleles. The small difference that separates our features and functions from a chimpanzee is even more impressive when it is understood that most of the two million alleles, in both humans and chimps, are inactive or neutral, leftovers from long past common genetic links. If we

---

discount all of the alleles that are common to all life, both plant and animal, the common percentage becomes larger and we are separated from our primate cousins by an even narrower margin.

There are more realistic and more significant DNA comparisons, not between our species and our closest animal relatives, but between all humans. In spite of our perception that humanity is divided into many divergent types and races, the genetic differences between all humans is much smaller than the differences found in most other species. As humans, we would expect our genetic makeup to be almost identical, but not as identical as it is.

In spite of our perceptions and prejudices, humans are genetically highly homologous, much more alike than is usual for most species. Some anthropologists postulate that our extreme genetic similarities are the result of our recent appearance on the evolutionary scene. Others attribute our unusually close genetic similarities to a population bottleneck created by a cataclysmic natural disaster that reduced the evolving human population to as little as 2,000 breading pairs 70,000 years ago. The natural disaster they assign as the cause of this near extinction is the massive Tobo eruption that took place on Sumatra in the late Pleistocene. The same eruption severely depopulated many other species and although arguments against the theory exist, there is substantial evidence to substantiate it.

Thick volcanic deposits from the, (Volcanic Explosive Index 8, mega colossal), eruption are found in sedimentary rock world wide, and as sediments on sea floors. More than 3,000 cubic kilometers of material were blasted into the atmosphere followed by long plumes of ash causing a six-year volcanic winter and a 1000-year-long instant Ice Age. Natural selection at the genetic level is inadequate to adjust to such abrupt changes in the environment, and there is evidence for many such events throughout geologic history where mass extinctions were followed by life in new forms. The Tobo eruption was one such event and the resulting environmental changes had a profound impact on the future of Homo sapiens.

After the Tobo disaster genetic options were narrowed to a point that natural selection could have easily eliminated the human genome, but humans were already equipped with the rudiments of cognitive awareness and those best equipped with the ability to reason had the ability to invent ways to survive. The survivors were few but shrewd and we probably owe our existence to their cunning and persistence. Genetics was unable to provide adaptations in a timely manor but memetic adaptations were fortunately available and produced immediate innovative behavior that saved the wise and weeded out the less aware. The only adaptive mechanism available for survival, other than genetic drift, is memetic adaptation and in this case, "the survival of the fittest" became the survival of the most cognizant and creative. Tobo, as a natural selection event narrowed the gene pool, widened the meme pool and focused both the genetic and memetic future of Homo sapiens toward cognizant, creative, adaptive survival abilities.

The Tobo scenario is based on both fact and speculation. As an accurate accounting of a critical point in human evolution, it has factual value in understanding how the environment can override genetic drift in evolution. As a deductive analogy, it has value in explaining how the natural selection of advanced states of awareness can result from memetic demands created by environmental stresses and how the culling of the human population can account for the unusual genetic similarities between all humans. Tobo also provides an example of how environmental conditions can focus natural selection. Cause and effect sometimes have a curious relationship. The fact that the preeminent life form on earth could be the result of a massive volcanic eruption may be one of these.

The impact of language on the development of human awareness

No real evidence exists for the advent and structure of the earliest languages used by man. Linguists who study *glottology*, (the origins of spoken language) differ greatly as to how and when it developed, but they do agree that language, (the ability to form concepts and communicate them), developed before speech, and that speech is only one form of communication. They also agree that *recursion*; the ability to imbed phrases within other phrases is a distinguishing characteristic of human speech and is only occasionally found within the capability of other species. Curiously, the ability is found in starlings. Even more interesting is the observation that only man seems to be able to use speech to ask a question. Even communicative primates who can answer complex questions almost always lack the ability to ask them. Human children, on the other hand, begin asking their first questions using question intonation, (voice inflection), long before they begin to use syntactic structures, This ability exists in all humans, without exception, regardless of culture or type of language, (be the language; tonal, non-tonal, into-national, or accented). The ability to question, along with the early development of question intonation appears to be embedded in humans in much the same way that instinctive behavior is embedded in other species. Unlike any other living species, we appear to be preprogrammed to be curious and ask questions.

Because of the social nature of early hominids, even before the advent of Homo sapiens, some form of interactive communication was certainly a part of the human repertoire. Speech developed as the primary form of communication after other forms were tested and after intellectual capacities and physical attributes, allowing speech had developed. Speech requires the mental capacity to embed concepts and references into an articulated sound, the motor skills and anatomy to create articulated sounds, and the auditory ability to receive and decode the vibrations after transmission. For speech to be possible, the human brain, the tongue, the larynx and the ear must all be adapted and ready for speech and must work in sync. To ask if the capability of the ear developed before the articulate tongue, or if the lowering of the larynx in the throat to allow vowel sounds was naturally selected because of enhanced speech capabilities, or because the resulting deeper voice attracted more mates, posits unanswerable questions. We need not answer these questions to realize that we are here, speaking, listening, reading, and writing, and that human speech developed, and that with speech, our awareness expanded.

Speech and advanced awareness are natural partners and compliment each other like a professional dance team that wins every competition. Deciding which partner, (language or awareness), is impressing the natural selection judges more and continually earning the genetic pass to the next level of competition, is futile. As mental capacity adds to language, awareness expands, and as communicative capabilities prompt genetics to add more synaptic pathways, awareness further expands, and the dance goes on.

In concert with man's advances in awareness, human language has developed from gesture to pictograms to verbalizations, and prompted additional advances in his awareness. Homo sapiens who embodied these advances survived and reproduced, and have become the dominant species on the planet. Neanderthals, Homo habilis and Homo erectus have been eliminated. Homo sapiens survive and thrive because of our advanced awareness. We encompass, in our individual synaptic repertoires much more information about our surroundings and ourselves, than any species. Our extreme informational base is due, in large part, to our language advantage and it has taken a quantum leap by adding to individual retentive and inventive capabilities, informational storage systems created by writing. We have jumped to a new level of evolution with new natural selection determinants. The digital language of Homo sapiens now surpasses the coded language of genetics in determining the future of life on Earth.

The impact of human awareness on the environment

There are very few humans alive today who are not acutely aware of the impact that Homo sapiens are having on the environment. The impact of our advanced state of awareness has made the Genesis account of Adam and Eve's apple eating experience quite relevant.

The Gods conferred and said; "Let them not eat of the fruit of knowledge least they become as one of us."

The story is of course a myth, but it is also a very prophetic and a powerful analogy. We have indeed become as one of them with god like powers, and we have not behaved responsibly. We have abdicated our responsibility not to overpopulate the planet leaving to fate, all decisions regarding reproduction. We have not taken responsibility for the other life forms on earth and are consuming them, usurping their habitats and driving them to extinction with our voracious unchecked appetites. We ignore the impact of our never-ending production of inventions and constructions as we modify the planet to fit our perceived needs and have polluted and imbalanced the biosphere. We will continue to be a threat to ourselves, and our planet until challenged by the results of our behavior, and are forced to change our perspective. Responding with environmental nudges, suggesting we balance the power of our awareness with natural selection, nature seems to be giving us advanced warning of impending hammer blows if we continue to ignore her basic arrangements. Unless we learn to cooperate, we face retaliation from an environment whose power far exceeds our civilized state. Unless we learn to cooperate with each other and with nature and to control the explosion of human reproduction, we face a de-selection by natural forces. It is now time to place your bets on the outcome of this match between man and nature! The odds are 100 to 1 in favor of nature.

Survival advantages of human awareness

Being human has definite advantages. Our advanced state of awareness, our advanced language, our cumulative knowledge, and our creativity, has given us control of natural processes and allowed us to create agriculture, great cities, and civilizations. Most humans now enjoy a stable existence insulated from predators and most small capricious environmental changes. The power of our advanced awareness gives us stability, comfort, long life and a survival advantage that minimize the effects of many natural disasters.

Survival disadvantages of human awareness

Our self-awareness is also inherently self-destructive. Our obsession with human preeminence has created perspectives that focus on dominance and away from our true place in the natural world. Before science refocused our attention on the world around us, human awareness was focused on topics related to human events and human behavior. Most ancient writings attest to this limited perspective, and the emphasis has been carried forward in religious, philosophical and historical writings.

The self-concerned awareness that accompanies our advanced cognitive abilities has created a culture of egocentric individualism that has diverted us from interacting realistically with our nurturing environment.

Our religions are based upon individual relationships to a god and are generally exclusive of group moral considerations or group decisions that go beyond restricted congregations. Modern political constructs are also based on exclusive groupings of individuals, languages and perspectives. Within these separated communities, accepted perspectives unite the group, maintaining order, and provide direction. Unfortunately, The same groupings divide humanity into arbitrary elements, which are often at odds with each other.

The capacity of these communal perspectives to provide a safe and stable existence for a community of individuals provides a temporary sense of security that becomes codified and protected.

Unfortunately, natural or self-imposed disasters create exigencies beyond the coping capability of these restricted religious and political constructs. The advanced state of human awareness is flawed by remaining focused on narrow and divisive perspectives.

CHAPTER THIRTEEN

Chapter Contents

CHAPTER THIRTEEN

ENHANCED AWARENESS

Tool Enhanced Awareness

Sensory enhancement tools

Man is separated from all other species by his capacity to make and use tools. A few other species use natural elements like twigs and stones as tools but only man shapes and invents tools for a variety of special purposes. The advent of the human species is marked in archeological records by the earliest use of tools. We have come a long way from flint cutting tools and stone hammers. The most significant tools invented by man have been those that augmented his sensory abilities.

We learn through our senses. We learn from what we see, what we hear, what we feel, and what we smell and taste. The tools we have invented to augment these abilities have allowed us to use our sensory synaptic learning abilities to explore the universe beyond the possibilities presented by our unaided senses.

Before the telescope, planets were errant stars and only speculative explanations existed for their strange motions in the sky. Before the invention of the microscope, reality smaller than a grain of sand was outside our observational capacity and reality was limited to the information our unaided eye could provide. Until we invented tools to see in infrared and with ex-rays, reality beyond the narrow visual wavelengths in the electromagnetic spectrum remained dark.

Sight is our primary sensory window to the world but until Galileo turned his small telescope toward the heavens in 1611 and until Anton Van Leeuwenhoek turned his convex lenses toward pieces of minute matter in 1670, our world was the same size as the world viewed by the many other creatures around us. The use of crystals and rounded glass to bend reflected light to create a rainbow and make things appear larger to the eye has been known since 1000 CE, and probably much earlier, but, not until 1590, when father and son eyeglass makers Zacharias and Hans Janssen experimented with multiple lenses, did the telescope become a useful tool. Not until 1690 when Leeuwenhoek melted glass fibers into small spheres did the microscope become a useful tool. These two sensory extending tools have taken our awareness from the limited view of immediate surroundings to the ability to see distant galaxies and the ability to see bacteria and the causes of disease.

To take our vision beyond the limited view offered through eyes evolved to see using only a sliver of the radiation flooding the universe we created tools to bring the remaining wavelengths of the electromagnetic spectrum into the realm of our sensory capabilities. The views we have created are of necessity secondary or transposed but allow us to see indirectly things invisible to the naked eye. By using film and wavelength conversion devices, we can now observe our surroundings in microwave wavelengths, x-ray, and infer-red.

We have also invented ways to excite atoms and molecules so they emit radiations we can record and review, and can now look at details inside a body or object without dissection. We also create images using sound waves to probe the Earths crust and mantle, as well as our own bodies.

The electron microscope, developed in 1931 by Ernst Ruska, has allowed us to see items near atomic size. Giant colliders allow us to explore the makeup of subatomic particles, and infrared and microwave telescopes allow us to look back in time and observe our universe as it was being born. These amazing tools allow us to see the reality around us beyond our evolved sensory limitations, and their impact on the perspectives of those who choose to look, is profound. Unfortunately, most turn away.

Memory enhancement tools

Some level of "event or situational memory" is evident in all living things. This can be genetic or learned, simple or complex but serves the same purpose as inanimate objects that retain mechanical memories. Memories carry information forward for latter reference or action. Memory in plants may be only a twisted tree trunk from an early injury or the budding of a flowering plant in response to a seasonal change, but these biomechanical responses are not as simple as they seem. Biomechanical responses are complex events which we oversimplify because we posses a special additional tool for memory enhancement called language.

The complex language of man has redirected his genetic selections and caused his neurological capacities to advance to extremes. Language, even in its most primitive form, is a powerful survival tool and an inducement to genetic advances in neurological and other physical attributes. In man, language has taken awareness and memory to levels far beyond other life forms and has had its impact compounded by his inventiveness and tool making abilities. One of the first enhancements man added to his innate capacity to speak and remember may have been the invention of numbers, and words for numbers.

Having a word to identify a mammoth was extremely valuable in coordinating a hunt. Having words that conveyed the number of mammoths would be even more valuable and probably developed quickly from sign language, the holding up of fingers to represent quantity. As a language's vocabulary grows, the memory capacity of those using the language grows proportionally. The connection between man's primary senses, sight, and hearing, is innate and hard wired in genetic synaptic connections. A predator or a mate recognized by image association can prompt an identical association in response to a predator's call or the mate's voice. This internal connection of sound and sight is essential for the development of language opened the door for the complex languages of man. We communicate both verbally and visually using a translation code to switch back and forth between these two sensory inputs and the translation process leaves neurological traces we call memory.

Except in rare cases, visual memory appears to dominate in man and can effectively increase one's retentive capacity when visual images are associated with names and numbers. Recall capacities are similar in all humans but extreme exceptions exist, and remain unexplained. The almost total recall ability of savants or the exceptional ability to recall and reproduce music by those with autism are probably related to genetic faults in synaptic sight sound connections.

The memory capacity of most mammals is huge. From birth to death during every waking moment, data is collected, stored, and recalled and individual retention capacities in mammals remained sufficient until language began to supplement the advantages of living in groups. The community of awareness of man was created by his early language skills, which led to the invention of sensory augmentation tools and an even larger meme pool. The volume of information produced by sensory augmentation tools, now far exceeds individual synaptic storage capacities. Without the emergence and development of a written language, this enormous amassing of information would not have been possible and could not have survived past a few generations. Any large library attests to the effect writing has contributed in establishing the world's largest pool of communal awareness.

With the introduction of the scientific method and the invention of more powerful investigative tools, the recording of information using only written accounts began to reach its limits. As the volume of information from discoveries became explosive, faster ways to store and recall the flood of information became imperative and getting beyond our naturally evolved, biologically based language required stepping beyond both genetic and memetic evolution. It required humans to harness one of the basic languages of nature; "the language of; "On or Off", "Yes or No", the same language used by genes in natural selection.

Learning to speak nature's digital language has allowed us to augment our capacity to gather and store information a thousand fold. Nature's larger processes are primarily analog but nature uses her "on or off" option at micro levels. Because the "one or zero", "on or off", "yes or no" method has application at the micro level, we have been able to create electronic tools to use and translate the digital language and to use digital devices to decipher and store even more of nature's secrets. Digital processing and recording has not increased the individual retentive capacities of the human brain, (not yet), but it has increased the retentive capacity of the human communal meme pool exponentially.

Tools that enhance reason and logic

Man has been inventing since he asked his first inflective question. He has been creating alternative methods and tools long before he began to use fire and lost his fur. At first he invented tools to supplement his hands, tools to smash, grind cut and kill, but as his prowess as a hunter-gatherer improved he began to pay more attention to subtleties in his surroundings and took a step beyond his innate animal instincts. By assessing the passage of time and by learning to predict the season's man improved his levels of comfort and survivability, and as his language grew to accommodate wider perspectives, he began to envision ways to track time with shadows and natural monuments. The movement of the stars, the sun, and the moon were constants that became even more important as he began to experiment with agriculture and animal husbandry.

The movements and patterns of the stars became important in his early cultural development and instilled a sense of regularity that became a tendency to recognize and use geometric shapes. Circles were the easiest to layout using any convenient piece of material as a radius, but other shapes followed quickly as early man discovered that knots placed at equal intervals in a vine or rope could become a measuring tool. From these early reasoning tools, man discovered that if he created three corners and wrapped the knotted vine around them, he could create a nearly unlimited variety of triangular shapes. Probably by accident, using a vine of twelve knotted equal lengths, a curious tribal member found that a vine wrapped around corners separated by three knots and four knots respectively, always resulted in a third side of five knots, and a square corner.

Reasoning tools, like power enhancing tools, are self-perpetuating. One tool leads to the next by suggesting an improvement. Having names for each individual finger may have led to the first numerical language and the birth of mathematics. Using crude measuring devices may have led to the birth of geometry. The use of stones and sticks to trace the movement of the sun and stars may have led to astronomy. Every observation led to a new question and a new invention. Man's reasoning tools have followed this self-advancing pattern and have allowed him to become prolific builders and great engineers. His reasoning tools have also given birth to science, mathematics, and logic. We have created a long succession of additional reasoning tools to supplement our individual reasoning capabilities and depend upon them to sustain civilization.

Before the advent of mechanical reasoning tools, like the abacus or the adding machine, we used language as a tool to enhance our reasoning abilities. By naming and symbolizing mathematical and logical patterns and putting them in visual forms that could be manipulated, we made great strides in advancing our abilities. The syllogism, Venn diagrams, arithmetic symbolisms, algebraic symbol manipulations, and the calculus, are examples that have augmented and improved the accuracy of our calculations. Electronic reasoning tools have taken us another step forward.

When electronic counting became available, punch cards took our capabilities forward by allowing the manipulation of larger amounts of data, and with the introduction of digital information storage, we took an even larger step. Circuitry miniaturization has taken us forward in our mechanical reasoning capabilities another power of ten and we can now collect, store, and analyze enormous amounts of information using powerful algorithms. Using our logic and analytical tools in concert, we have increased our interpretive and reasoning capabilities a thousand fold and we are still inventing. If the increase in computing capacity continues at its present pace, we should be capable of creating the storage capacity of the human brain in just a few decades, but being able to mimic the nuance and adaptive ability of the mind as created by four billion years of natural selection, may always be beyond our reach.

Tools that enhance social awareness

The true power of humanity lies in its collective awareness. The knowledge base of humanity far exceeds any individual capacity and is dependent upon the constant support and maintenance of individuals. Only a continuing, collective, and cooperative community of mind can sustain civilization. Awareness of ourselves as individuals initiated our rise above other life forms, but it is our ability to subjugate ourselves to societal orders, to learn and contribute specialized skills, to train and educate our young, and to replace and enhance essential individual contributions, that makes our technologically based society possible. Being cooperative is at the root of all moral and legal codes. Knowing that cooperation is essential requires a sense of others and common perspectives.

Spoken languages precipitate social awareness. Written languages expand social awareness, and mass media distributes social awareness, creating world-views and social interactions that are altering perspectives and demand cooperative efforts on even larger scales. The tools that have altered our social perspectives and individual responsibilities are primarily technological, and have been introduced without considering the consequences. The ability to share discoveries, inventions and information adds enormous power to humanity as a whole, demands extreme cooperation and has produced a global economy, but we remain emotionally tribal and competitive in our basic individual perspectives. Our evolved tendency to gather in groups with similar physical characteristics, a common language, and common beliefs, continues to divide us into small groups incompatible with our more encompassing current cooperative state. The persistence of our tribal instincts creates challenges and conflicts. This fragile balance between cooperation and competition is also one between our advanced state of communal awareness and primitive primal individual tendencies, and is expressed most strongly by the differing perspectives created by secular science and religious dogmas.

Paranoia and distrust created by race, language and culture is easier to overcome than the deep differences that exist between secular and religious perspectives. Resolving this conflict in perspectives has become humanities greatest challenge and the outcome will determine the future of life on earth for centuries to come.

Social awareness expanded beyond gossip and the town crier with the invention of the printing press. Sacred scrolls containing edicts and proclamations read only by priests were made available to the masses, and the ability to read became essential for anyone who wanted to continue participating in society. The ability to purview lofty diatribes by the common people quickly changed the tone of such edicts and announcements and the languages in which they were written. The printing press also created immediate access to current events, influenced methods of governance, the behavior of social groups, and economic conditions. The balance between self-interest and societal interests is altered by every new tool that enhances social media. Increases in social media also impose a greater responsibility on the individual to be reliably informed.

The telegraph played a major role in the outcome of the Civil War and in the settling of the West. The telephone allowed the masses to communicate with each other instantly over great distances and allowed commerce to take place at a much faster pace. The radio added another dimension to the dissemination of social information, and television augmented this ability further. Individual humans are now aware, almost instantly of happenings around the world, but the tools of dissemination still have their limitations. Radio and television require funding, and those funding the broadcasts influence the style and content of the information provided. Commercial broadcasts require sponsors and prudent sponsors pick-and-choose programs that appeal to their demographic markets. Government sponsored broadcasts are filtered for different reasons, some laudable, some not. Even public sponsored broadcasts, dependent upon contributions, pander to the tastes of their contributors, and privately funded broadcasts, both non-profit and for profit, have well defined agendas.

Social information and perspectives change over time because of these intentional and unintentional filters in ways difficult to measure and society adapts, without realizing it, as broadcasters present options to social perspectives. We are blind to the consequences of most evolving social dynamics because we are not capable of forecasting or controlling them.

Our latest tools, created by computer networking and satellite communications have increased the pace by which information can be disseminated and has added a new dimension to our ability to communicate with each other individually and between groups. The electronic methodology that creates these networks also dilutes privacy and exposes information intended to be private to unknown parties. The mass of information and great numbers of self-appointed disseminators, reliable and unreliable, add to the confusion and create an innate distrust of information and fundamental societal sustaining principles.

The Tower of Babble as depicted in the Bible is a good analogy. The new social media may create a better mix of cooperative self and societal behavior or it may undermine successful long-term organizational experiments. When everyone in a room is talking at once you can only recognize bits and pieces of relevant conversations, you have to shout to be heard, and have trouble concentrating. Being able to make an intelligent decision on anything requiring a serious discussion in such a situation is doubtful.

Awareness Retention Through Instruction

Informal guidance

Beyond random observation and mimicry, retentive minds in many species carry patterns of learned behavior forward to future generations through subtle promptings. Not all individuals in a migratory species need to possess the navigation skills to reach a migratory destination if some possess the skill and take the lead. Migratory followers may or may not learn the route or be able to recognize stop over resting sites and might not even feel the urge to migrate. The guidance given by those who are instinctive and have learned the route is, however, of value to all, and extends the value of their individual awareness to all the individuals in the group. In primitive species, informal guidance is the primary form of expanding group awareness and of preparing the young for survival. The advantages of increasing states of awareness have continued to be naturally selected. The advantages of improved methods of memetic transfer also evolved because of the advantages they provide.

Intentional guidance and instruction

The self-perpetuating loop, between advanced awareness prompting adaptive behavior, and advancements in behavior prompting increases in awareness, creates a receptive synaptic capacity in many species that is open to intentional instruction. Memes transferred by intentional instruction between parent and offspring are the most common and easiest to observe. Birds teach their chicks to fly by example, and by forcing them out of the nest or, in the case of ground nesting birds, by repeated takeoff runs and much coaxing. Intentional instructions are the primary method for meme transfers essential to survival in advanced life forms. Predator species teach hunting techniques, while their prey teaches escape and evasion techniques. In man, language has formalized meme transfer and taken instruction to advanced levels. Man's survival is now completely dependent upon his formalized educational processes.

Formal education

Only man, using his curiosity, creativity, and sophisticated language has been able to accelerate his own evolution beyond natural processes. He has done this by creating a communal pool of awareness through millions of cooperative investigations, and by creating engineering skills, amazing tools, and complex industrial, economic and social organizations. To support and maintain these complex cooperative efforts requires specialization, organization and continuous education. Collectively these great works are called civilization, but civilization like other ephemeral states, is not static, requires a constant effort to maintain, and is subject to the caprices of nature.

Henry David Thoreau took note of the fragile structure of civilization long before it became today's mega mechanical complex. He literally stepped back into nature to see if he could get a glimpse of the future, and his comments from "On Walden Pond", have become especially germane. We now find ourselves trapped by the awesome responsibility of constantly servicing every detail of our complex creations. We have organized our food chains, our sheltered comforts and our ability to communicate and coordinate our activities around electronic and mechanical devices, have built great structures to house our essential records and numerous activities, and have populated the land in such a way that individual transportation capsules and great paved ways are essential for a continuing commerce.

What was once the simple skill of a blacksmith forging shoes for horses, has become a complex industry of automobile design and maintenance requiring robots and thousands of learned technical skills. What was once the simple task of digging a well and installing a bucket on a rope to retrieve water, has become a complex industry designing and maintaining great dams, sophisticated water treatment plants, pumping stations, and miles of viaducts and pipes. The oil lamp has become international electrical grids. The wood fireplace has become distribution networks of natural gas that span continents. The animal and human energy that once powered early civilizations has been replaced by tapping into the remains of buried and decayed living forms that accumulated over millions of years and will be depleted it in just a few hundred years.

Thoreau's analogy of a greedy man reaching through an opening in a fence to gather manufactured goods only to find that with his arms full, and unwilling to let go, could never get back through the opening. The analogy is especially relevant today. We are stuck! We all have our heads stuck through the fence and with no way to get back are forced to use our brains or perish. We either cooperate to survive in the situation we have created, or watch our great civilizations crumble around us. We cannot rely on ancient texts or old methods. There are too many of us to move back to Walden Pond and most of us would not survive even if it were possible. We have no choice but to put forth the effort needed to maintain what we have created and improve it. We are temporal beings, and if we want our children to survive, we must teach them the millions of skills needed, and they in turn, must teach their children. Extensive formal education is no longer an option. It is a survival necessity.

CHAPTER FOURTEEN

Chapter Contents

CHAPTER FOURTEEN

ALTERED STATES OF AWARENESS

Inadequate awareness

Anything short of omnificence, the full awareness of all things and events past present and future, can be construed to be a limited view of one's surroundings, but omnificence is an idealized fiction. Usually attributed to gods, omnificence is beyond proof or reasonable inquiry, and arguments concerning the reality of such a state, and by inference the limited value of lesser aware states, hinge upon empty postulates. A more pragmatic perspective in the assessment of awareness is to view it as the result of a slow process of awakening through a slow and persistent evolution of matter, life, and an emergent property called awareness.

There appears to be no awareness in inanimate matter. In contrast, awareness appears to be a universal element in living matter and has a common history and common elements that have diversified as life evolved. Each species and individual has its own unique perspective. The aware perspective of an earthworm as it burrows into moist soil in search of decaying vegetable matter is dependent primarily on a passive level of awareness and its perspective is very different than the aware perspective of a coyote as stalks its prey in an open field.

To capture the essential nutrients it needs for sustenance, the coyote uses passive level instincts in conjunction with an active awareness of its surroundings directed by cognitive assessments of tactics. Is the awareness of the earthworm insignificant in comparison to the awareness of the coyote, or only different, and which is a better assessment of reality? The answer is; that both are equal in their assessment and more importantly, they are equally relevant.

Relevance is a practical measure of a living forms activity to insure its continuance using various categories and attributes of awareness. Without awareness, a living form cannot respond to its environment and without relevant responses, a living-form, along with its awareness, will cease to exist. Relevant awareness is the naturally selected match of a living thing to its environment. When a relevant awareness is distorted by injury or a genetic, gemetic, or memetic fault, the new match of the aware form's activity to its environment is altered. A distorted awareness is rarely productive. Death is only a living form with its awareness removed.

A coyote with an injured eye has its awareness distorted by the alteration of its visual sensory inputs. A coyote with a genetically malformed eye can have a similar distortion. A coyote with a brain injury that affects its visual cortex in such a way as to restrict the interpretation of visual signals also has a similar distorted awareness. A coyote with an injury or a genetic anomaly that alters the coyote's instinctive connective neurology, or a coyote with a restricted ability to make cognitive decisions, all suffer from a detrimental distortion of their aware state. All of the examples given pertain to physical alterations affecting the creature's awareness and are the result of either injury, genetic or gemetic faults but there are others.

Memetic elements can also alter awareness. Memes are transferred as observational or instructional bits of information that become part of a living forms synaptic repertoire and faulty memes can distort awareness. Memetic transfer becomes increasingly important as species depend more on acquired synaptic arrangements and less upon genetically implanted arrangements.

In man, memetic transfer has become essential. A completely unschooled human has little chance for a normal life.

The Coyote analogy above highlights the complex nature of awareness and its complex interface with both its physical host and the environment. Because of its fragile genetic, gemetic and memetic connections, awareness is susceptible to distortions. Man is especially susceptible to distortions in his awareness from false memetic inputs, (lies and exaggerations). A clever orator can mislead an entire civilization by repeatedly transferring false memes to large groups of naïve listeners. Shielded from valuable memetic transfers or being subjected to false memetic transfers creates inadequacies in awareness that can have negative impacts on group and individuals. Natural selection however will eventually override any lie.

Suppressed awareness

An aware state requires energy. In the case of advanced mammalian awareness, quite a bit of energy. The energy to sustain awareness comes from the metabolic processes of each physical host and each hosted awareness waxes and wanes as the host adjusts its metabolic rate. Passive awareness, in primitive life forms, adjusts its intensity as the host adjusts its growth rate and chemical activities with day and night cycles, with the seasons, and with shifts in wind or sea currents. More active states of awareness, usually associated with more mobile life forms, begins to exhibit rest periods associated with the host's metabolic patterns. In creatures with advanced cognitive states of awareness, periods of nearly complete shut down seem to be required. Without these periods of inactivity, awareness looses its stability and begins to become unproductive and eventually counterproductive. Regular rest periods for aware states have been naturally selected as essential for reasons we do not yet understand.

Suppressed aware states can also be self-induced by intense periods of focus on one's body or on intense mental or physical activity. When a cognitive creature concentrates with intensity on a single activity or event, it closes off other aspects of its awareness and loose track of other activities, even its surroundings. From these disassociated periods of awareness, humans have created the idea of a subconscious and invented the early fields of psychology. It is also from our ability to self induce these dissociated states that transcendental meditation has developed, religious revelations have been produced, and creative moments occur.

One of the most productive inventors of the twentieth century, Thomas Edison, used a special technique to access these creative moments. When stumped by a problem, he would sit in his rocking chair and hold a spoon loosely in one hand. He would concentrate on the problem and at the same time let himself drift into sleep. The moment he transitioned from a relaxed state to a sleep state his hand relaxed and the spoon would drop awakening him. Often he found the solution he was seeking now implanted in his aware state. Edison was balancing his states of awareness to access information that his cognizant, aware state was masking. Many others have referred to these creative semi aware states including Tesla and Einstein.

Advanced life forms selectively balance their levels of awareness to fit immediate situations. Monkeys resting as a group concentrate on grooming and interactive group behaviors that reinforce their social bonds. During these periods, they call upon their active and instinctive levels of awareness and suppress their cognitive aware states. They are enjoying each-others company and searching for ticks, they are not problem solving or escaping from a predator.

The changing activity levels of individual awareness are as natural as the decision to sit walk or run. Man has also learned to induce partially suppressed aware states and completely suppressed aware states using chemicals. Modern medicine, biological research, and veterinary medicine, are all facilitated by our ability to use these chemical suppressants. Invasive medical procedures would be nearly impossible without them.

Self altered awareness

Creatures with advanced cognitive awareness seem to delight in temporarily altering their awareness. Animals will willingly drink fermented beverages and apparently enjoy the temporary euphoria caused by the alcohol and return for more. They will also breath fumes from chemical springs and eat leaves that produce psychedelic effects. Humans have taken the art of self-altering their own awareness to extremes, and are constantly inventing new ways to bend their awareness into new forms. Chemically altering their state of awareness is the preferred method for humans to distort the reality around them. This ability is also our best evidence for the chemical basis of awareness.

Altered states that suppress cognizance and elevate states of awareness that are more primitive, heighten synaptic connections between instinctive/emotional hard wiring areas of the brain and centers for active awareness. Enhancing these synaptic connections beyond their normal chemical reactive limits, exaggerates these connections and can cause drug dependencies and permanently self altered conscious states.

The use of drugs to alter one's awareness is not new. The oracles of Delphi resided in a temple built over hot springs whose fumes induced a state of euphoria. The guidance the oracles gave sounded inspired but was really just drunken prattle. Opium has been around since the poppy was first cultivated. American Indians reveled in chewing peyote and the Incas smoked many weeds. Today's drug culture is endemic and concentrates on the more addictive drugs to create repeat customers who will go to great lengths to calm their exaggerated synaptic connections. Supplying those addicted to self-altering aware states with drugs has become the most profitable business on earth.

Alterations by genetic drift, fault, or injury

All individual awareness is dependent upon the physical health of its supporting sensory and neurological systems. Helen Keller's neurological system was intact and healthy, but was shut off from her immediate surroundings by nonfunctional visual and auditory sensory organs. The power and impact of language on the human condition is nowhere better illustrated than in her use of language to compensate for her sensory deprivations. Using only the sense of touch to establish contact with the world and fill her nearly empty synaptic neurological connections with substitutes for images and sounds, Helen replaced auditory and visual inputs with touch and tactile symbols to described things she could never see or hear. Once the tactile language was learned, she could communicate with other humans and her nearly empty synaptic base was quickly filled. Without the miracle of language, Helen Keller would have led a dark and silent life. Without the miracle of language Homo sapiens would have become extinct or be just another common primate.

The natural selection of mutually reinforcing attributes is common. Advanced sensory abilities combined with advanced neurological capacity and a capacity for language is the trio of talents natural selection has chosen for the advanced aware state of man. As a cooperating community of awareness, man has become godlike. As individuals, our mental capabilities remain totally dependant upon the physical sensory and neurological systems that produce them. Genetic faults that affect this ability evidence themselves as syndromes and mental disorders that are similar and recognizable. Very few of these maladjustments are capable of correction by therapy or treatment beyond minor adaptive results.

Levels and attributes of awareness are not static. States of awareness drift and are resorted and retested by natural selection, just as genetic traits are resorted and retested as their genetic coding drifts. The process seems essential to evolving better adaptive living forms and better adaptive states of awareness. Injuries, illnesses and severe genetic anomalies produce altered states of awareness in individuals that do not often carry forward to a next generation. Natural selection may play a passive role in eliminating individuals prone to a particular injury or illness or by shortening or eliminating their reproductive opportunities. Subtle and small genetic coding errors, on the other hand, can produce altered physical traits that can be a positive, negative or neutral in influencing viability.

States of awareness are more susceptible to immediate alterations, temporary and permanent, than are physical traits because they are dependant upon both genetic and a memetic influences, and because they rely totally upon the health of a physical host for energy and activity. Awareness has evolved with physical forms to a state where, in advanced forms, it can supersede genetic influences but it will always be dependent upon the physical health and activities of a living host.

CHAPTER FIFTEEN

Chapter Contents

CHAPTER FIFTEEN

NEW PERSPECTIVES

A curious condition

I am aware, that I am aware of writing about my own awareness. I am also aware that others around me seem to sense their own awareness in a similar way. How did this strange loop in self-awareness come about? We all share this same curious condition. What awakened life from a few self-replicating chemicals? How did inanimate matter reach a point of complexity capable of taking a first glimpse of its surroundings, and what process propelled this first living spark to a point where living energy began to analyze itself'. From the time man became self-aware he has asked these question and for most of his history has answered them with fanciful stories.

Numerous gods and myths have been invented to satisfy man's obsession with his self-awareness, (sometimes called his soul), and his imaginative creations have shaped his perspectives, his societies, and his religions. This single persistent quest, looping from awareness, to self-awareness, to self-analysis, has done more to shape mankind's past, (and now his future), than any other. This single universal compulsion to understand our own awareness, and explain what happens when it ends, remains dominant in most of man's endeavors.

Seeking answers from nature by using scientific methods is a new approach to this search and by cooperating in our search we have learned a great deal about our surroundings, and how circumstance has shaped our physical form, but we are still not certain about the origins of our awareness. Unlike the ancient Egyptians, who thought the purpose of the brain was to produce mucous, we recognize the brain and its attached sensory organs as the source of our awareness but are still mystified as to its origins and its qualities. In part, our confusion is the result of our unwillingness to accept our awakened state as a natural condition, a condition shared by every living thing from bacteria to buffalo, a condition as natural as hair or fingernails, not strange at all.

Awareness, as a universal attribute of all life, is marvelous but not miraculous. We have come to accept the great diversity of living forms as the product of a natural process, a slow adaptation of physical attributes and function to insure survival in ever changing conditions. We call the process *natural selection* but it is primarily a process of elimination. Environmental conditions suitable for a specific living form inevitably change and eliminate previously well-suited life forms. New forms better suited to the new environment replace those eliminated because as life replicates, it allows for small deviations from generation to generation. Forms that continue in new conditions are those with chance deviations that fit the new environment. We would find the process more satisfying if there was evidence of a guiding hand, instead of chance, or if we could identify some goal for the process, but there is no evidence for either.

Awareness was only one of several attributes inherent in the earliest of living things but it; like the necessity for a membrane to separate its internal chemistry from its surroundings; the ability to gather nutrients and produce energy; and the ability to replicate, was an essential attribute. These four essential attributes, have been carried forward through every replication and every environmental test and remain essential attributes of every current living thing. From this perspective awareness, as an ability to monitor internal and external conditions is basic, normal, and extremely relevant. Awareness is not strange. It is essential.

Living things have developed complexity as a quality that allows more rapid deviations from a previous generation and thus a better statistical chance to avoid elimination as environments change. The four essential attributes of life, including awareness, have also developed in complexity. Cell membranes have gathered to become bark, skin and shells. Adsorbing nutrients through cell walls has become respiration and digestion. Simple cell division has become the production of haploid cells for enhanced gene swapping, and simple chemical awareness has become complex sensory organs and neural centers for processing sensory information. Our self-awareness is basic, normal, and essential for our survival.

We are the product of exploding stars. We are made of stardust and our physical form and functions are the product of four billion years of adaptations to earth's changing environments. Life's serendipitous beginnings may have occurred in a comets tail or in a tidal pool. We will probably never know, but we cannot deny our awareness or the awareness present in every other living thing, and can expect to find the same essential quality in every form of life we may yet discover in the ocean's depths or on other planets. Awareness is more than mind, more than intelligence, more than consciousness, more than reason, and more than cognizance. It is all of these, but more importantly, awareness is an essential component of the creative process, and is common to all life

<u>Cognizant limitations</u>

From the perspectives created by scientific inquiry, life appears to be only one of many naturally occurring events in the continuing evolution of the universe. The book of creation is open for us to read but requires translation from the all-inclusive language of nature to our simplified language of symbols and analogies. In the dimension of physical realities, there is an apparent flow of order evolving out of disorder, but our limited conceptual capabilities, even when enhanced by artificial modes of intelligence, fall well short of being able to grasp this flow in its entirety. Because of our limitations, we examine nature's patterns and processes one small piece at a time. The patterns we observe may approach reality but will always remain limited by our inability to grasp the total picture. We have no choice but to explore within the limits of our observational and mental capabilities, and within the limits of our language.

The symbolic language of mathematics has proven to be our most useful tool for manipulating and reducing nature's methods to understandable sets and patterns. Having developed the ability to symbolize quantities, relationships, dependencies, causality, and many other elements of nature, we apply the rules of our mathematical and logical symbolism to ferret out further details by thinking beyond our innate capabilities. A complex equation, when reduced to simpler more understandable terms by applying strict rules for symbolic manipulation, can be very useful in reducing an unfathomable natural process to a concept we can begin to understand. We have created symbolic repetitive thought patterns to get beyond our conceptual limitations and to allow further investigations into the physical universe around us. Computers have further enhanced our capabilities by allowing us to store and manipulate large amounts of information well beyond our ingrained capabilities. We have also invented tools for greatly enhancing our basic sensory capabilities and are no longer restricted to what the unaided eye or ear can detect. We have come so far in our exploratory capabilities that it would now be impossible to explain our expanded perspectives to the wisest of ancient prophets or philosophers. We have truly disconnected from our past and see the universe around us in extremes far beyond the capabilities of our ancient forbearers.

A new view of morality

Evolution has established a perpetual and ever changing balance of living forms and aware states. This balance is complicated by its many interface activities as species compete for space and nutrients. Some involve symbiotic support systems, some ruthless competition, some empathy, and some tolerance.

Natural selection operates without favoritism or pity and, other than insuring continuance, has no advanced plan and is indifferent as to what form life takes. The huge number of extinct species attests to this indifference and the large number of living species attests to its persistence. The moral core of natural selection, if it exists, would probably be "make it work no matter what."

Compassion, empathy, anger and aggressiveness are behavioral traits that humans assign to life as it cooperates and competes to keep the biosphere stable. In fact, these traits are eliminated or modified with indifference by natural selection and contribute little. Nature can be cruel or caring or both to insure the continuance of life, and uses genetic drift to test various forms, instinctive behaviors, and various forms of awareness to insure successful adaptations.

The accepted concept of human morality assumes that an established set of standards for behavior is required. It assumes that these standards are preexisting, or god given, and are implicit and obvious. The concept also assumes that compliance with these standards will result in societal order, a proper fit with natural surroundings, and an end to excessive competitive behavior. Any difficulties encountered with social arrangements, in dealing with natural surroundings, or with aggressive behavior, are assumed the direct result of non-compliance with these revealed moral standards.

Unfortunately, the success rate of various moral codes and standards has been abysmal. Humanity has flourished, not because of a compassionate benevolence, but because of a ruthless rape of natural surroundings, social orders based upon hate and competition, and self-aggrandizing perspectives that are completely out of sync with natural selection's more balanced approach.

Natural selection insures life's continuance through a careful mix of cooperation and competition and humanities success may be temporary unless we reconcile our perspectives to nature's natural processes. The book of nature is the history of creation, the history of the universe, and the history of life on our planet. These histories include the patterns that have made our advanced awareness possible. We still see and understand only a small part of the reality around us, and our best chance for survival is to investigate further, correctly interpret the patterns of nature, and comply.

Humans are blessed, or cursed, with a high level of self-aware cognizance that has produced self-aggrandizing perspectives. Instead of seeing ourselves as simply the latest test in an ongoing experiment, we see ourselves as the final attainment of a creative objective, standing separate from the natural processes around us. To avoid being relegated to the status of a natural being, we invent spirit dimensions that exist beyond sensory capabilities. To overcome the obvious affinities we share with all living creatures, we create imagined relationships with invisible omnipotent powers and create afterlife promises. Morality is simple, if the rules are spelled-out, and administered by an observing and intervening super-being. Morality becomes much more complex, and open to interpretation, if it is the product of natural causes. We feel compelled to simplify and codify our morality because we are the most successful and dangerous of predators. We do not trust our own instincts, or each other.

If we accept our advanced state of awareness as the latest unfolding of a natural process, we must also begin to accept responsibilities beyond the commandments of unseen gods. We must accept responsibilities that are implicit in our ability to redirect natural selection. Nature's rules are discovered only through trial and error and are enforced by survival challenges, not by rewards or punishment in an afterlife. Our advanced cognitive awareness has elevated us to a level where choice overrides simpler methods for the natural selection of compliant behavior.

As the universe awakens in us, we assume responsibility for its future and become the decision maker for the future of our planet. If we fail, others, here or elsewhere, will continue in the awakening process. If we fail, others will continue to explore and learn and will participate in the continued awakening. If we succeed we will bring with us the morality of correct choices that allows us to survive and will have made our greatest contribution. The process of natural selection will persist in either case, and will continue to give preference to those aware forms with ingrained behavior patterns that are well adapted to the reality around them.

Morality in man is currently based upon writings stating acceptable and unacceptable behaviors appropriate to the time of the writing Sacred texts are static, but societal perspectives, like the environment change, and morality, like living forms, must adapt to remain relevant..

Moral codes are not unique to man. Most species with a developed communal awareness depend upon some level of collaborative compliance by its individual members. Generation after generation survive by protecting and passing on the gemes and memes that allow the group to function successfully. Each new hatchling and each newborn is infused by instinct and instruction with the group's behavioral (moral) standards. These behavioral compliance standards are different for every species and every group, and prove their advantage in the same way that the earliest congregations of single celled life proved their advantage by initiating multi-cellular life.

Cooperative behavior codes have been naturally selected, not because they are ordained, but because they improve survivability. Without forethought, life flows into every available ecological niche both as individuals and as cohesive groups. The morality of each group also flows into these niches for testing by natural selection.

In groups of individuals, possessing more advanced aware states, complex-sorting rules become ingrained as they naturally select their own standards for dominance and position. These selections are made through competition, display, or actual conflict. In species with a Cognizant awareness, as in primates and man, birthright also plays a part in these selections. The development of these complex group behavioral patterns can be traced back along their developmental paths to genes becoming gemes, and memes becoming themes, a significant unit of transfer for survival information.

From genetic coding for advanced neurological states, to genetic encoding for neurological systems prewired for reflex and response, to mimicry and instruction, group behavior has been naturally selected in the same way that body types have been naturally selected to fit existing conditions. The rudiments of our morality are the result of natural selection. Our cognizance and creativity have added the rest and the results of this mating of naturally selected group behavior patterns, and cognizant impositions is the wellspring for all of our moral perspectives. If we really want to know where we are on this journey toward a natural accounting of our morality, we need to retrace our steps through our history and through the traces of natural selection that have led us to our present state.

Morality has its roots in early group behavior and early group rules remained simple until advanced states of active awareness complicated things. As more advanced levels of awareness were naturally selected, more varieties of social groupings became possible and dominance and social standing within groups became more important. These arrangements are everywhere in nature and can be observed in pods, flocks, herds, swarms, schools and especially in the organizations of man. Understanding that the morality of man is only an advanced form of evolved behavior common in nature, should be an easily acceptable concept, but very early in human development, tribal rules were elevated to mystical proportions and wrapped in ritual. Accepting human moral standards as a combination of naturally selected behavioral standards and common sense demystifies human morality and runs contrary to our religious traditions and most holy writ. Having the rules for human behavior exposed as naturally evolved patterns contradicts religious claims for man's special relationship with God and deeply ingrained beliefs cannot be removed without damaging civilization. Finding a middle ground between nature's revelations and religious revelations is therefore essential to the future of humankind. The established sets of moral codes that form the foundation for our laws and behavioral edicts have come primarily from religions. To remain effective and continue to be the underpinning of social orders religious interpretations must remain open to expanding perspectives.

The scientific method is a neutral examination process without moral content. It is not the enemy of religious beliefs, and although religion may never be a part of the processes we use to examine the natural world around us, we need not conclude that discoveries made by science cannot become a part of our religious perspectives.

Religious leaders steeped in both religious tenants and science understand that our quest to understand nature is also a quest to find an underlying purpose hidden in the processes that have formed the stars, planets, life and mankind. These few, bi-polar faithful scientists may be our best hope to open our moral senses to the additional responsibilities that new discoveries have placed upon us. Scientific investigations are a recent human endeavor and they will continue. The application of scientific discoveries sustains all modern civilizations. The long history of religious introspective discovery also continues, but instead of welcoming the expanded view of the wonders around us as a gift of nature, scientific discoveries are viewed as interference in god's will and advocate a "hands-off" approach to controlling natural processes and a rejection of a self-directed future.

Religious and scientific differences

Religious beliefs are as old as humanity. Science, on the other hand, is barely four hundred years old. Religion is an established and static set of beliefs that has created most societal perspectives. Science, on the other hand, is a method for inquiring into the workings of the observable world around us that creates an ever-changing pool of information and alters societal perspectives. One static, one fluid, one defining, one redefining, one based upon the unexplained (miracles), the other based upon explanations offered up by nature, one with head down and eyes closed, one with head up and eyes open. It is not surprising that science and religion have difficulty in finding common ground. Since the birth of the scientific method, science has unintentionally challenged religious beliefs with its discoveries and caused religions to become defensive. Before science, churches could speak openly about the wonders of nature, now they step into an unfamiliar secular realm and avoid it. Science also avoids stepping into religious realms by avoiding even a hint of purpose within their findings. Religion and science have turned their backs on each other declaring that the other's approach produces a dangerous and false vision of reality. The animosity that separates these two worldviews masks the fact that both are inextricably woven into the fabric of civilization and neither can be removed without causing a complete collapse.

Religion's appeal is evidenced by dozens of competing variations, billions of worshipers and hundreds of thousands of buildings designated as places of worship. Science's appeal is evidenced by the proliferation of new technology produced by the application of scientific discoveries. We drive to church in machines made possible by science and then pretend they don't exist as we participate in ritual worship. We give thanks for food produced by scientific methods and search out hospital chapels in order to pray while doctors use science to cure or save a loved one. Religion and science have a curious relationship. Like rival siblings, they continually squabble. They are forced to coexist because they belong to the same family. Both belong to the human family, but they are very different. Understanding those differences may help mollify the conflict.

Religions are established belief systems, usually based upon ancient texts and individual revelations that explain mankind's place and purpose in the universe. They are ancient perspectives sustained and perpetuated by evolved rituals and regular worship.

Almost all religions support a belief in an observing god that intervenes in worldly events and human affairs and listens to individual prayers. Most also support a belief in an after life that is programmed to reward or punish according to one's behavior during this life. Religions are static and resist changes but they too evolve over time as they branch into many different forms. Religions rely upon belief rather than investigation or evidence, and often engage in confrontation as a way to protect their belief systems. Religions use sacred texts to refute both other religions and scientific findings, and usually do not engage in activities that produce new knowledge or discoveries outside the boundaries of their beliefs. Strict religions confront science and technologically based perspectives to undermine their credibility whenever possible.

Science, on the other hand, is not an established belief system. Science is a method of investigation. Science does not search for answers concerning man's place or purpose but investigates natural occurrences searching for clues and patterns that will help explain them. Scientific discoveries are the result of group efforts not individual revelations. Even the ideas of Copernicus, Newton, Darwin and Einstein were based upon the findings of previous investigators and required verification by many other investigators before being accepted as temporary useful explanations. Science is not static. Scientific theories are not at all like tenants of faith. Religious tenants are accepted as final and absolute. Scientific theories are assumed temporary until better theories replace them. Science evolves rapidly as new discoveries are made. It relies upon evidence found in nature and rejects unfounded explanations.

Science is a prodigious producer of information that leads to technological advances and new medical techniques.

Science, in its pure form, is not immoral. It is however, amoral. Scientists search into nature's methods and offer their discoveries to others who use the information as they please. Nearly always scientific discoveries enhance human dominance over nature and provide the basic information used by inventors to produce innovations in agriculture, technology and medicine. In most cases, these innovations improve the human condition, but not always. Scientists rarely ponder the future effects of their discoveries. In this sense, science is blind. It digs deep into nature's methods and freely exhibits whatever it discovers. Scientists follow the excitement of discovery wherever it leads and offer up knowledge that is quickly converted to applications that can be very beneficial or extremely destructive.

Religion, in its purest forms, is also not moral, it, like science, is amoral. Religions combine basic tenants of belief, based on ancient texts and individual revelations to create ethical behavior codes and formats for ritualistic worship. Religious behavioral codes usually have relevance to harmonious human relationships but are primarily guidebooks for entry into an after life. These codes, if interpreted literally, supersede both logic and the laws of the state and are not open to question. Their strict applications, like the applications of scientific discoveries, can be beneficial or extremely destructive. History holds many examples of the destructive effects of both misplaced applications of scientific discoveries and religious overzealousness.

As a determinant for the future of human kind, a serious evaluation of the effects of natural selection on human behavior is not a part of science or religion. Both ignore it. There is no common ground between the perspectives created by the process of investigative discovery and divine revelations because natural investigators rarely look away from their experiments and the faithful rarely look up from their sacred texts. Both will agree that we are responsible for our own future, but even this initial agreement is subverted by religious expectations for an Armageddon that makes serious attention to any group behavior modifications in this life, moot.

Scientists are secular in their approach but not necessarily atheistic in their beliefs. Scientific information is in a constant state of revision as new discoveries are made. Science is many books being written and rewritten. Science looks outward to find truth in physical surroundings. Science is the expression of a naturally selected state of awareness in which the universe examines itself and awakens with every discovery. There is no larger miracle than this, and to be aware of ourselves as the instrument of this awakening, whether chosen or randomly evolved, places us in a preeminent position as the primary, naturally selected instrument of the awakening of inanimate matter. On this planet we have become responsible for our own adaptations to the environment, and because of the power of our awareness, we have also become responsible for the environment. We face an inevitable choice; accept our responsibilities for the environment and our own future or abdicate and let chance determine the future. If we abdicate, we need to put science on hold and undo much of technology. The abdication of our responsibilities to nature is what religious zealots advocate when they encourage acts of terrorism, when they preach Armageddon, or when they fight to exclude science from the classroom. 'Let god's will be done' equates to; "Give it up people", this life isn't worth understanding or fighting for."

Some religious leaders understand the havoc and horrors that an actualization of these teachings will produce and in some perverse way, many welcome them. Others are naive to the implications, others put on (god's will) blinders, and others put the importance of their privileged place within the church above all else. Religion is not inherently evil and secularism does not hold all the answers. The general population goes on oblivious, caught up with concerns for immediate needs as they are swept along by the great tide of natural selection. Individuals of faith and those with secular perspectives all follow and take advantage of technological advancements and attend church seemingly oblivious to the dichotomy until a sensitive issue arises. The ability of general populations to accommodate the mix of religion and science by being pragmatic and reasonable persist, until an issue arises that attracts extremists.

If religion and science are ever to coexist peacefully, and they must if we are to survive, they must find common ground and this can best be accomplished by illuminating the key difference that separates them. We insist on defining this difference by asking the question. "Do you believe in God"? We then insist on a simple yes or no answer. For the unquestioningly faithful, the answer is easy and any hesitation in answering is interpreted as a lack of commitment and therefore a "No". However, any reasonable person will hesitate if the question is asked by someone of another religion or by someone whose beliefs you hold in question. Hidden in the question, when asked this way, are two words with many definitions, "belief" and "god". Is Allah the same god as the Lord of Hosts?" and how do I know what belief entails? Is belief blind obedience, appropriate emotional responses, ritual, an understanding of holy writ, or a commitment to a church? If religion is ever to accommodate scientific information, it first needs to become tolerant of other religious beliefs, to teach their tenants as well as their own, and to begin to understand and accommodate each other

Religion and science both influence societal order and a mix of reason and beliefs form the foundations of all societies. These formalized mixes become the laws that govern citizens and the rules they must follow. The variety of governmental forms resulting from these random and ever changing combinations are themselves grand experiments, each presenting new forms for testing by natural selection as they compete and measure themselves against reality. Understanding the many perspectives that mingle in a formed society helps to explain the society's laws, their internal control mechanisms, their attitudes, and their reactions to other formed societies. These are very complex arrangements, often beyond any accurate analysis, but at the core of each is an observable balance between reasoned and religious tenants.

Rules for individual behavior are usually specific in despotic governments or within strict religions, and flow from divine revelations, like the proclamations of a King, Mohammad or Moses. In contrast, societal rules derived from secular perspectives are implied from reasoned accounts of our observable position in an observable universe. In secular perspectives, the "Thou-Shall-Not" attitudes of Christianity and Islam are supplemented with "You Should" statements to insure human compliance with natural selection and these are usually nonspecific mandates.

Religious mandates promise rewards in an after life, Secular mandates promise rewards in this life. Are these promises realistic? Are they mutually exclusive? The questions we ask determine the answers we find, but answers also shape the next question.

We are searching earnestly for life elsewhere in the expanding universe. Are we alone and if so why the wasted space extending out billions of light years in all directions? If we find extraterrestrial life, will it be similar to life on Earth? What happens to our egocentric assumptions regarding our relationship to God if life is found on other planets? How will we classify our extraterrestrial finds?

We are discovering planets at an astounding rate and we need to prepare for what appears to be an inevitable encounter. Alien life may defy physical descriptions but even alien life will posses a level of awareness, a pace of awareness and a scope of awareness. From these traits, a way to establish a comparison exists and hope for a mutual understanding is possible. To make the comparison, whom will we send to greet our first extraterrestrial visitors? Will we send scientists, ambassadors, warriors or missionaries, and whom will they send to greet us?

All truth goes through three stages:
First, it is ridiculed.
Then it is violently opposed.
Finally, it is accepted as self-evident.
Schopenhauer

ABOUT THE AUTHOR

Vern Arthur Westfall is a philosopher, a pilot, a teacher, and a designer of fine homes. He holds a bachelor's degree in philosophy from Miami University and attended the United States Air Force Academy. He has flown many aircraft, including jet tankers and supersonic spy planes. He has also served as a foreign liaison officer, has extensive experience in civil engineering and a strong background in architecture. His writings include fiction, non fiction, short stories and poetry.

TOPICS

www.ingramcontent.com/pod-product-compliance
Lightning Source LLC
Chambersburg PA
CBHW020745180526
45163CB00001B/359